How the World Listens

How the World Listens explores our everyday and professional interactions with sound. The book aims to uncover the human relationship with sound across the world and to reveal practical ways in which a better understanding of listening can help us in our daily lives.

This book asks how sound is perceived, expressed and interacted with in both remarkably similar and dramatically different ways across the world. Using findings from a new scientific study, conducted exclusively for this book, we embark on a globe-trotting adventure across more than thirty countries, through exclusive interviews with more than fifty individuals from all walks of life, from acousticians and film composers to human resource managers and costumiers.

How the World Listens is essential reading for anyone with an interest in human relationships with sound, including but not limited to sound design and music composition professionals, teachers and researchers.

Tom A. Garner is a Senior Lecturer in Interactive Technologies at the University of Portsmouth. Originally from a background in popular music performance, Tom now works extensively in emergent approaches to auditory perception and extended reality (XR) technology. His previous books include *Sonic Virtuality* and *Echoes of Other Worlds*.

Sound Design
Series Editor: Michael Filimowicz

The Sound Design series takes a comprehensive and multidisciplinary view of the field of sound design across linear, interactive, and embedded media and design contexts. Today's sound designers might work in film and video, installation and performance, auditory displays and interface design, electroacoustic composition and software applications, and beyond. These forms and practices continuously cross-pollinate and produce an ever-changing array of technologies and techniques for audiences and users, which the series aims to represent and foster.

Sound Inventions
Selected Articles from Experimental Musical Instruments
Edited by Bart Hopkin and Sudhu Tewari

Doing Research in Sound Design
Edited by Michael Filimowicz

Sound for Moving Pictures
The Four Sound Areas
Neil Hillman

Composing Audiovisually
Perspectives on Audiovisual Practices and Relationships
Louise Harris

Designing Interactions for Music and Sound
Edited by Michael Filimowicz

How the World Listens
The Human Relationship with Sound across the World
Tom A. Garner

For more information about this series, please visit: www.routledge.com/Sound-Design/book-series/SDS

How the World Listens

The Human Relationship with
Sound across the World

Tom A. Garner

Routledge
Taylor & Francis Group

LONDON AND NEW YORK

Cover image: Aha-Soft/Shutterstock.com

First published 2023
by Routledge
4 Park Square, Milton Park, Abingdon, Oxon OX14 4RN

and by Routledge
605 Third Avenue, New York, NY 10158

Routledge is an imprint of the Taylor & Francis Group, an informa business

© 2023 Tom Garner

British Library Cataloguing-in-Publication Data
A catalogue record for this book is available from the British Library

Library of Congress Cataloging-in-Publication Data
A catalog record has been requested for this book

ISBN: 978-1-032-01453-1 (hbk)
ISBN: 978-1-032-01566-8 (pbk)
ISBN: 978-1-003-17870-5 (ebk)

DOI: 10.4324/9781003178705

Typeset in Bembo
by codeMantra

For James (cheers for the seagull) and Sophie.
  Dedicated to R.M. Schafer.
  Sorry for all the times I spelt your name wrong.

Contents

Figures

Tables

Acknowledgements

A special thank you to everyone who took the time to give their wonderful insights for this book. And to my anonymous contributors, sorry I cannot acknowledge you here, but know you're very much appreciated:

Diletta Acuti
Elisée Amegassi
Iñaki Aparicio
Mahmood Asgari
Amit Barde
Miloš Batoćanin
Lucia de la Bella
Franz Bierschwale
Borneo Brown
Brenda "Meni" Bulnes
Michael Lee Cullen
Kurt Jonathan Engert
Angélica Estevez
Andreina Gomez
Dana Hamdan
Sam Hughes
Heikki Kossi
Daniel León
Nicolas Liebing
Brecht De Man
Marc Sander Montant
Muhammad Muhammady
Mehrdad Nickbakhsh
Tim Ostanin
Gareth Owen
Miles Partridge
Arturo Pineda
Hatim Qadeer
Si Qiao

Daniela Saadeh
Smita Sahu
Can Sarac
Patricia Sunandar
Rasmus Brun Thorup
Laksha Thulasitharan
Shoko Tipping
Mark Trusov
Jahangir Uddin
Anna Vaernes
Daniel Villamizar
Wendy Wang
Kristina Wanyonyi
Adele Wong
Alyson Watts
Jon Weinel
Ruiyi Yang

Introduction

Palmerstone Road, Southsea. 24th of May 2021. Late afternoon

We start with the most characteristic of Pompey introductions, the incessant cawing of the seagull. The wind creates a gentle, low rumble as it passes by, interrupted by the sudden searing blare of a workman's drill. It's over in a moment. In the distance, there's a muddy mix of human voices, the most noticeable, children's screams. These screams don't come across as particularly emotionally laden. No overt indication of fear or joy, more a neutral expression that, to my middle-aged brain, suggests all young people's voices naturally operate at a greater decibel level. An obnoxious decibel level one might say, but maybe that's because I'm walking alone and these are the youthful voices of people in social interaction, which is perhaps emphasising my solitude – the miserable, melancholic sod, meandering down the street and scowling at any expression of positive emotion.

This is a book about sound. A little more specifically, this is a book about our relationship with sound. It is about how we perceive and make sense of sound, how we extract information from it, and how we attribute information to it. It is about how we move, act, behave, and interact with sound. It is about how it makes us feel and think, how we make use of it, and how we make, express, and craft it. This work presents a collection of thoughts, feelings, and impressions from over 120 hours of interviews, collected over a period of 5 months, with a grand total of 50 contributors, and representing just shy of 35 countries across the world. Within these discussions, we explored the human relationship with sound. We talked about what they understood sound to mean to them and how it featured in their everyday lives, in their home, their commute to work, and their workplaces. Together, we considered the social and professional aspects of sound, as well as what they believed 'good' and 'bad' sound to be. Responses from the contributors revealed many insights and perspectives on sound that, despite being a researcher in the

DOI: 10.4324/9781003178705-1

discipline for over a decade, I found to be genuinely surprising. They made me question my own relationship with sound and consider ways in which this relationship could be 'nurtured' – possibly even providing me with tangible lifestyle benefits. I do hope that you find the details within to be as surprising, challenging, and beneficial as I did.

Origins and precedent

As with many academic ideas before this one, the first seed that eventually became this book was planted down the pub. To be fair, that's really my northern vernacular cutting through as it actually started on a little patio of a bar in Prague. There myself, my PhD supervisor at the time and a delegate from Germany who we had met at the conference we were attending were sat, cheerily chugging Czech lager, talking mostly nonsense. Exactly how we got onto the topic I cannot recall, but the conversation moved to how so many German words did not have direct English equivalents, despite identifying concepts most of us would be familiar with. Several, such as *wanderlust* and *schadenfreude*, we have effectively adopted into our everyday language, but many have not. Some of my particular favourite examples include *torschlusspanik* (a sense of time and opportunities slipping away relative to one's age), *sturmfrei* (having the freedom of the house to yourself in your parent's or partner's absence) and *dreikäsehoch* (directly translating to 'three cheese high' and used as a reference to small children). As the conversation continued, I posed a two-part question that set in to motion the thinking that would shape this book:

- I wonder how many words out there, across all the languages of the world, express sound but have no direct equivalent in English?
- Could some of these words reveal hidden concepts about sound and could knowing more of these words improve someone's understanding of what sound is and what it means?

This was, admittedly, quite a few years ago but the idea remained lodged in my head, sometimes hidden away, sometimes looping around my inner monologue much like an earworm. Funnily enough, after relenting to the idea one afternoon, I decided to do some light investigating to see if I could find even a single example of non-English words relevant to sound with no direct English equivalent and found the word *earworm*. In its more recent history, the meaning of earworm rather than the word itself has migrated from Germany to find common usage in English. Both the German and English terms are decidedly similar (earworm and *ohrwurm*), possibly due to another usage that was also shared between countries, referring to earwigs and certain pests. The full story can be read on the Merriam-Webster site[1], but roughly between the 1950s and 1980s, the German ohrwurm and the concept it represented did not exist in the English vernacular. Of course, the phenomenon of what is typically referred to in academic literature as 'Involuntary Musical Imagery (INMI)' would still have been experienced across England before the 1980s, with the 1968 Beatles' track *Ob-La Di, Ob-La-Da* featuring as a well-recognised earworm in a recent INMI study (McCullough & Margulis 2015). So period, poor-suffering Brits were grappling with the acute, if distinctly first-world, problem of incessantly looping musical phrases stuck in their heads with not even a word to describe their plight.

Now, not to give the whole game away, but those two questions posed above ended up on the cutting room floor of this project. A few remnants make their way in, scattered throughout the pages, but in the end, the function of this line of thought was to inspire much wider thinking on how the human relationship with sound differs around the world, and I'm grateful to those initial ideas for the impact they had. Sorry, you had to be sacrificed. Something else came up.

The habitus

Applied to sound studies by Sterne (2003), *the habitus* is a concept created by French sociologist Pierre Bourdieu (1972). In broad, less eloquent terms than Bourdieu himself would likely use, the habitus is the theory that the sociocultural grounding of our personal reality subconsciously engenders all of our perceptions, thoughts, and behaviours. The effect of the habitus can be thought of as history becoming nature (Bourdieu 2005). Focussing the notion of the habitus upon sound, Sterne argues that such sociocultural factors (including custom, style, and social outlook) have a profound effect on how we listen and, consequently, on how we communicate both through sound and about sound.

In a revealing opening section of *Remapping Sound Studies* (Steingo & Sykes 2019), Gavin Steingo revisits a moment whilst travelling between Sandton and Soweto, South Africa. Steingo was struck by not only the dramatic soundscape shift between the two locales but also the unexpected cultural cause. As Steingo explains, drivers in Sandton are encouraged to keep the windows of their vehicles closed to avoid the hassle from road vendors. As a result, the sonic interior of the moving car is isolated, controlled, and personalised. In Soweto by contrast, closed windows are indicative of mistrusting the local townspeople, conversely making such drivers themselves the subject of mistrust. In Soweto, pedestrians and vehicles are consequently in constant intimate proximity. As Steingo describes it "…a car's boundaries become porous [and] intimacies are established through sonic exchanges" (p.39). Steingo proceeds to discuss two key points relevant to non-musical sound within sound studies theory. The first is that technology (and not just audio or music technology) is heavily contributing towards an increasingly isolated sonic experience. The second is that listening is associated with biopolitical investment and efficiency, which, as Steingo explains, concerns circumstances in which aspects of an organisational structure (this could be a company but equally a whole industry or even an entire country or global initiative) can be changed to control the sonic effects of that structure upon its workforce. Purposes for such sonic changes include reducing accidents and promoting the health of workers, improving general productivity, or simply reducing the number of compensation claims being submitted. These notions return us to the idea of *the habitus* by observing how numerous and longstanding points of great difference between cultures of the world can have a profound and tacit effect on the sonic environment – a sonic environment in which people of that culture are deeply and persistently immersed and within which they carry out their daily lives.

In addition to matters such as technology, organisational structure, and localised cultural factors, other noted sound studies theorists have attested to the relevance of the habitus with regards to professional practice. Bijsterveld (2019), for instance, posits that factors linked to the functions, routines, and workplace environments of different professions can have powerful effects on our relationships with sound in both conscious and sub-conscious ways. This seems relatively self-evident when considering creative professions such as musicians, voice actors, and sound designers, or professions with characteristically noisy environments such as construction. However, as we shall discuss a little later, unique profession-based interactions with sound can extend to a much broader range of careers.

So, if we are to accept the theory of the habitus and its potential effects on sound, plus the observation that the phonetic construction of language differs significantly across the world, there is (I hope you'll agree) a decent justification to feel emboldened at the possibility of investigating this further and actually yielding some interesting results with meaningful practical application.

Sociocultural theory

Vygotsky's renowned work on sociocultural theory (see Lantolf 2000) broadly divides into four components which, in line with the theory, influence our cognitive development as children and consequently the 'shape' of our adult minds and how we think, feel, interpret, and communicate. These components are: (1) social, historical, and cultural mediation; (2) thought and language; (3) social interaction; and (4) method of instruction. Here, connections between sociocultural theory and the habitus become almost obvious in terms of the meaning and expression of sound. We can only express our thoughts upon sound linguistically with the tools our language affords us. We cannot experience sound outside our own personal timeline or cultural space, inescapably drawing in history and society as we interpret meaning from what we hear. We cannot separate our experience of sound from task or instruction and, as social creatures, we possess little power to extract our experience from our relationships. Indeed, much of the rationale for this book finds roots in Vygotsky's work, but where he did not expressly focus on specific sensory modalities, we are considering the sociocultural effects as they pertain predominantly to sound.

In related work, Jónsdóttir and colleagues (2015) present a layered model of sociocultural theory that centralises the individual and wraps them in the *community*. This community comprises dominant caregiver influence alongside relatively weaker influence from schools, social services, religion, media, politics, and the nation. This dominance of caregiver influence is made clear in a remarkable 2010 study by Kärtner and colleagues that investigated *contingent responsiveness* – the method by which parents and caregivers, when actively responding to their infant children, will instinctively follow a particular temporal response pattern that their infant needs in order to formulate foundational social interaction. A caregiver needs to respond to an infant vocalisation within a so-called, 'latency window' (different studies argue over the precise duration of this window but it is somewhere between one and nine seconds). If the response is presented after this window, the infant will not connect the response to their call and the opportunity is missed to reinforce their sense of sonic agency upon the world. The purpose of the Kärtner study was to identify differences between sociocultural context upon the nature of contingent responsiveness, specifically under what circumstances a caregiver would choose to respond, and what their response would be. More than 150 families were observed in this study, across six countries: Germany, China, India, Cameroon, and the USA. The results revealed that sound was, by a substantial margin in most cases, the primary modality for both infant activity and caregiver response, except for the rural Cameroon sample, which more evenly balanced sound and touch. The results of the study also revealed the caregiver's desire to encourage either autonomy or interdependence (a reciprocal sense of balanced dependence between two or more people) as a strong correlation to their contingent responses. The more a caregiver aspired to engender independence in their infant child, the more their contingent responses were delivered using sound. By contrast, caregivers wishing to inspire more interdependence would rely more evenly on touch and sight responses, alongside sound. Kärtner and colleagues theorised that this was due to a difference in physical distance between response modalities, with touch being proximal (*at* the infant), sight being intermediate (also known as medial) and sound being distal (*at* the caregiver).

This starts to join all kinds of dots together as we consider the, admittedly messy, web of interconnected issues that constitute the meaning and expression of sound across sociocultural lines. As we shall return to later, sound is largely interpreted across the world as distal, meaning at the object of its source. This connects to Kärtner's caregivers relying on it more heavily to interact with their children to whom they wish to instil greater independence as it denotes a greater sense of distance. It is also worth noting that Kärtner and colleagues (2010) also consider prior research that shows such distal responses do indeed engender greater levels of autonomy, meaning that even at a few months old, the human relationship with sound meaning and expression already has some significant definition. It also suggests that the path to building that relationship begins to diverge, based on our culture, from almost the moment we are born. But how far does this rabbit hole go? How dramatically different is the human relationship with sound across the world? Are these differences meaningful? How do they affect us as we develop, and how do they shape the individuals we become?

Hopefully, the above serves as a convincing precedent for the expedition we are about to embark upon, throughout which we shall address matters of meaning, culture, place, everyday experience, and professional practice, all in relation to sound. We begin, however, with a wee bit of groundwork.

Note

1 https://www.merriam-webster.com/words-at-play/earworm-meaning-origin (accessed 12.05.2021).

References

Bijsterveld, K. (2019). *Sonic Skills: Listening for Knowledge in Science, Medicine and Engineering (1920s–Present)*. London: Springer Nature.

Bourdieu, P. (1972). Les stratégies matrimoniales dans le système de reproduction. In *Annales. Histoire, Sciences Sociales*, 27(4–5), 1105–1127.

Bourdieu, P. (2005). Structures and the Habitus. In G. M. Spiegel (ed.) *Practicing History: New Directions in Historical Writing after the Linguistic Turn* (pp. 72–95). New York: Psychology Press.

Jónsdóttir, S., Guðjónsdóttir, H., Jónsdóttir, S. R., & Gísladóttir, K. R. (2015). Creating meaningful learning opportunities online. Bank Street Occasional Paper Series 34. http://www.bankstreet.edu/occasional-paper-series

Kärtner, J., Keller, H., & Yovsi, R. D. (2010). Mother–Infant Interaction During the First 3 Months: The Emergence of Culture-Specific Contingency Patterns. *Child Development*, 81(2), 540–554.

Lantolf, J. P. (2000). Introducing Sociocultural Theory. *Sociocultural Theory and Second Language Learning*, 1, 1–26.

McCullough Campbell, S., & Margulis, E. H. (2015). Catching an Earworm Through Movement. *Journal of New Music Research*, 44(4), 347–358.

Steingo, G., & Sykes, J. (Eds.). (2019). *Remapping Sound Studies*. Durham: Duke University Press.

Sterne, J. (2003). *The Audible Past: Cultural Origins of Sound Reproduction*. Durham: Duke University Press.

1   Sound and theory

Approaching Southsea Castle. 17th of June 2021. Mid-afternoon

As I pass the main seafront road and approach Southsea Common, numerous point sounds spring to my attention. Over my head, the whirs of a light aircraft sound. The wind, which until this point has been a low rumble, now creates a much more scattered tone as it whistles through the many trees. Even a solitary wood pigeon can clearly be heard, cooing away. In the distance are the gentler and more reverberant chirps of small birds. The cars on the road remain the most constant background sound, rising and falling in what feels very similar to waves on a beach. Very occasionally, I notice the sound of my own footsteps, but only when walking on a slightly gravellier surface. The pavement feels eerily silent unless I shuffle my feet. The soundscape feels very still, and steadily the frothing of waves lapping at the beach becomes apparent. The sound is unmistakeable and at once familiar but also novel. Living a few streets from the beach and hearing that sound truly emphasises how little I visit the seafront, despite knowing how positive the sensory experience is. Maybe I'm trying to conserve the novelty, as if hearing the scattered frequencies as the water hits the pebbles will somehow lose its value if I hear it too often.

Asking the right questions

Thank you for joining me on this journey. I'll be honest – at the time of writing the very first draft of this chapter, I had no idea how the book itself would turn out. I was sat at my desk at home, with no clear recipe, and with what felt like only the vaguest idea of what I wanted to write. This section was initially written in the present tense and was in effect, a confessional, from which I confronted the dawning reality of the obscure task ahead. A lovely bit of tension to kick-start the writing process. It is now a year later. I am a

DOI: 10.4324/9781003178705-2

couple of weeks away from submitting the draft to the publisher and trying to refine some sense into my account of what has transpired.

Let us start at the beginning. All good research starts with a good question. With reference to a 1959 piece by American sociologist Robert K. Merton, Hammersley and Atkinson point out that "[...] finding the right question to ask is more important, and sometimes more difficult, than answering it" (2007: p.25). It often doesn't help that most research questions are themselves a culmination of many smaller, feeder questions, meaning that to find a good question usually means you must find *several*.

What is a good question? A good question is, quite simply, one that is worth answering. Otherwise, it is without purpose. A purposeful question of this very moment would be *"what determines whether a question is worth answering?"* As the seminars on attaining research funding will probably tell you, this is a matter of impact, and impact is a matter of the effect that answering your questions will have upon the world. This is measured broadly across three dimensions. The first is scale. Affecting a large number of people (what academese would refer to as 'stakeholders') is better than a small number. The second is meaningfulness. Having an effect that the stakeholders perceive as considerable, substantial, or genuinely meaningful is good. Changing the lives of every human on earth but in a way that no one cares about, or even notices – that's not so good. The third is time. An answer that promotes lasting change has greater value than one promoting a change that almost immediately disappears.

Identifying the stakeholder is the easy bit. It's you, the reader. That makes my first measure of success straightforward – reach a lot of readers. But what about the meaningful effect? What questions on sound would a reader be intrigued by? Going back to that time of writing my first draft of this chapter, and despite having worked in the field of auditory perception and sound design for over a decade, I didn't know the answer to this either. And so, I began the process of investigating the field to get an updated awareness of the current status quo of sound as a discipline, exploring some of the current edges of knowledge, the challenges in the field, the nature of sound within a wider context of human experience, and looking for a few things that we simply don't yet know. The hope being that this would help build an understanding that would then reveal questions that were genuinely worth asking.

To dispense with the false dramatic tension, I already said that this book is about exploring the human relationship with sound in the introduction, so the *big* question you already know. However, what this question specifically means – what the feeder questions are, who would benefit from knowing the answers to these questions, and in what way they would benefit – all of that currently remains unclear. Provided I haven't put you off and you're willing to join me on this little pre-expedition, this chapter is effectively about us getting to know sound. A quest to find a few good questions.

A (quite) brief review of sound theory and perspectives

For many people, myself included back when I was a wide-eyed first-year PhD student, the idea that the meaning of sound was contentious within both the academic and popular communities was somewhat surprising. Often compared to its visual sibling which can be both static, say in a photograph, or dynamic as in film, sound can only be experienced dynamically across time. As Elvira Di Bona observes, "[o]ne of the striking aspects that characterizes auditory experience is [its] temporal nature" (2017: p.107). These temporal features inescapably portray sound as a linear phenomenon, but sounds can also be layered. What was one sound could become two sounds that, depending on their nature, could be perceptually merged so that, to the listener, there is now just one sound. Is the sound of a car engine just that – the sound of a car engine? Or is it the culmination of multiple valve and piston sounds? What if there are two cars? Is it the sound of a car engine, plus the sound of a car engine that I am hearing, or is it the sound of car *engines*? If that made your head hurt a little, the temporal, organisational, and cumulative natures of sound are all important matters, but they are just the tip of the iceberg. For this reason, I feel that there is a strong argument for our first question to be a foundational one: what is sound?

What is sound?

On a decidedly old webpage (particularly for an online media source), *The Guardian*[1] poses an ethical conundrum: *if a tree falls in the woods and there's no one there to hear it, does it make a sound?* At the time of writing (it may still be accepting responses if you fancy contributing…), a grand total of 46 answers were posted on this forum, 25 of which were based in the UK. Admittedly this is not the result of a robust analysis, but a broad consideration of the posts reveals two things. Firstly, this online forum, like many online forums, is subject to people getting angry over a seemingly innocuous question (or ignoring the question entirely to go on a rambling philosophical detour). Secondly, there was an even split between respondents who clearly and confidently stated that the tree would indeed make a sound and those who equally confidently posited that it would not.

So, which is it? Does the tree make a sound? Does that depend on other definitions being set first? What is the mechanism of sound? Is it purely physical or is it also perceptual? If it can be perceptual, can sound be entirely perceptual in some circumstances? Is any of this debate meaningful and does understanding sound from a specific viewpoint bring any practical advantages? This section isn't presented as a definitive answer but will instead review some of the key academic texts and scientific studies that have sought to provide such answers over the years. In doing so, the aim of this review is to reveal the various points of contention in the meaning and expression of sound, that will serve as the foundation for later stages in our expedition. Ultimately, we're not seeking to know if a tree falling in the woods does indeed make a sound. We're seeking to know if people from different walks of life across the world answer differently, and why that might be.

Sound as a soundwave

Known by many names, from the 'acoustic definition' or the 'mechanical perspective' to the 'standard view' or the 'physical perspective' (we'll stick with 'acoustic definition' here), understanding sound as a soundwave remains undeniably common in both popular and scientific understanding. It is often referred to as the 'common sense view of sound'. It remains the dictionary definition of sound and represents the majority (albeit slim) in the theoretical material I pass on to my sound design undergraduates. The acoustic definition explains sound fundamentally as variations in air pressure. These variations are compressions (describing the pressure pushing molecules closer together) and rarefactions (meaning a release of pressure causing the expansion of molecules). A soundwave is characterised by the two key properties of frequency and amplitude. Frequency identifies the number of complete wave cycles per second. It is measured in Hertz and is perceived as pitch. The greater the frequency, the higher the perceived pitch. Amplitude describes what is known as 'maximum particle displacement'; this is simply the difference between the greatest compression point and the greatest rarefaction points of a soundwave cycle. Amplitude is perceived as loudness, the greater the amplitude, the louder the sound.

The origin of a soundwave is typically described as an 'excitation', denoting some form of material vibration that causes adjacent molecules to become displaced. The displacement of molecules, how a soundwave travels, is known as 'propagation'. One commonly occurring point of confusion arises when trying to conceptualise how a soundwave propagates through a medium and in the exact way in which it displaces molecules. Almost all diagrams visualising a soundwave present the wave as transverse, or two-dimensional: meaning the wave travels along a relative forwards plane whilst disrupting molecules along the vertical plane. Strictly speaking, this is not accurate, but visualising a soundwave in this way makes it significantly easier to conceptualise and measure. In reality, soundwaves do travel (or propagate) through the environment in a relative forward motion but typically disrupt molecules either one-dimensionally or three-dimensionally, depending upon the medium. If the medium is gaseous, such as through the air, a soundwave travels longitudinally (or one-dimensionally), compressing and expanding the molecules along the same vector as the soundwave is moving. If the soundwave travels through either solids or liquids, it will disrupt the molecules in all directions.

Before a soundwave can reach our ears, it will first propagate through the environment (also known as the propagating space) where it will be subject to the acoustic effects of reflection, reverberation, diffraction, and refraction. Within this propagating space, reflection occurs when the soundwave bounces off a reflective surface to produce an echo. This is similar to reverberation, but the key difference is that echo describes only the reflected signal, whilst reverberation describes the combination of all reflections and the original source. Two other acoustic affects worth knowing about are diffraction and refraction. Diffraction describes a change in soundwave direction as it passes through an opening or interacts with a barrier. Refraction is also a change in direction, but one caused as a soundwave passes through a different medium (such as water). Both diffraction and refraction commonly affect the soundwave and its dynamics. Soundwaves also travel faster through liquid because of the more densely packed molecules within the medium. Because of this, the displacements will travel a greater distance whilst losing energy at roughly the same rate as they would in the air. The upshot being, you would expect a soundwave to sound notably louder if both you and the source were submerged in water, compared to in the open air, even if the source and the distance stayed the same. It would, however, also be expected to sound rather muffled, as liquid typically filters out some of the higher frequencies. The same so-called 'low-pass filter' effect is heard when a soundwave travels through a door, as the material reflects and deflects a lot of the higher frequencies, again creating a muffled sound.

The acoustic definition explains sound across three stages. The first two, excitation and propagation, we have just discussed in brief. The third stage is 'reception'. Beginning at the pinnae (or outer ears), auditory reception describes the soundwave's movement through the ear canal, to the middle and then inner ear, where it is translated into electrical impulses which travel along the auditory nerve to the cortex. The shape of the pinnae helps to funnel soundwaves arriving from different angles effectively into the ear canal. The precise shape of the pinnae and their configuration (as in, their exact positioning relative to the head and body) influence the incoming soundwave's acoustic quality. Like fingerprints, pinnae are unique to the individual, meaning even within the acoustic definition of sound, we all physically hear the world in (at least) slightly different ways.

The middle ear comprises the ossicles and ear drum. The ossicles are a set of three particularly small bones, commonly known as the hammer, anvil, and stirrup due to their unique shapes. Their role is to convert, or transduce, incoming soundwaves into the mechanical vibrations that they were before they became soundwaves. Next, the ear drum connects to the ossicles and is the barrier between the middle and inner ear. Its function is to amplify the vibrations from the ossicles and convert them into fluid-propagated vibrations, the required signal format for the next stage in the pathway, the inner ear. The inner ear houses the cochlear, a spiralling cavity along the length of which is the basilar membrane. Atop the basilar membrane sits the Organ of Corti, a four-row arrangement of hair-like cells, called stereocilia. Whilst it is helpful to understand the complete structural anatomy of the ear, it is the function of the stereocilia that is most critical to understanding how we receive sound. As the fluid-based vibrations transduced by the ear drum travel along the basilar membrane, they create a so-called 'shearing force' that stimulates the stereocilia. The more intense this force, the greater the electrical impulse generated and the greater the perceived loudness of the sound. Different sections of stereocilia are responsive to different frequencies, depending upon their placement along the basilar membrane. This enables our inner ear to discern between different frequencies, enabling us to perceive pitch.

Whilst the stereocilia provides us with essential pitch and loudness information, our body has one further trick up its sleeve. Through the gift of two ears, the auditory pathway enables us to localise (identify the relative position of) the source of a soundwave, based on the differences in soundwave intensity and moment of hearing between the ears. These two localisation features are known, respectively, as 'interaural intensity difference' and 'interaural time difference'. Interaural intensity difference describes the way in which the stereocilia of one inner ear can be stimulated more than the other, which we would normally expect to result from the soundwaves possessing more amplitude at the point of reaching one ear, compared to the other. If for instance, a dog was positioned to our right-hand side and began to bark, the soundwaves from the barking would likely be louder in the right ear because they travelled directly from source to ear, whilst those received by the left ear would be either transduced through the body or reflected from nearby surfaces, either way significantly losing amplitude energy in the process. As a result, the stereocilia of the right ear would be more intensely 'sheared', generating more electrical activity from the right side which the brain can readily interpret as "the barking dog is to my right".

Interaural time difference is the same principle, but it is the point in time at which the soundwaves hit the ear that is different. For example, soundwaves from our barking dog if received first at the right ear, then 600 milliseconds later at the left would be perceived as almost entirely to the listener's right. Any delay between the ears of up to 600 milliseconds can still give an impression of left or right position, with our sensitivity to the specific scale of the difference helping us to localise a sound source across a full 360° with impressive accuracy (2–3.5° in optimal circumstances – see Freigang et al. 2015). Much like the eye, the ear is a genuinely miraculous achievement of life; its physiology brilliantly corresponds to the physical components of the sensory phenomenon we need it to receive.

The acoustic definition is certainly rooted in fact and physical principles that do not engender much in the way of questioning. The definition does, however, provoke debate, specifically with regard to its boundaries. Even as the acoustic definition extends to matters of transduction and auditory processing, it does so in a highly mechanical and reductionist manner that some may argue misses the vital richness and nuance of the human experience of sound. It also places the emphasis for understanding and communicating sound very much on the mechanical properties and processes of soundwaves and the physical characteristics of the auditory pathway. As we shall return to later, this is not inherently a bad thing. It merely presents us, as do all the perspectives we shall discuss, with some benefits and some limitations. How we understand sound in specific terms provides a focus that enables very real progress in specific areas of scientific understanding, and in the development of practical solutions, but may have limited application elsewhere. The important thing to appreciate is that there are multiple perspectives on sound, each asserting its own points of focus and, in turn, affording its own value if properly applied. In line with the ethos of this book, the wider the array of perspectives we know, the more solutions we are open to. Here, we consider some of the most prominent alternatives to the acoustic definition.

Sound as an object or an event (or both [...] or neither)

For many of us, sound as an object makes the most sense when we consider our everyday experiences. If, for example, you heard the sound of a desk fan whirring and I were to ask you to identify that sound, you would most likely respond, *"it's a fan"*. Presuming this was your response, it is fair to extend from this to suggest that you had perceived the sound as an object. Furthermore, you perceptually placed the sound *at* the object. We could extend this inductive reasoning a step further to argue that you identified the sound as an object, precisely because you placed the sound at that object. As Di Bona (2019) argues, understanding where sounds are enables us to know what they are. This brings us to our second fundamental question: where is sound?

Convoluting the question of *what is sound* with *where is sound* is commonplace within the scientific literature. In many cases, it is easy to see why. Referring again to Kärtner and colleagues' (2010) study outlined in the introduction, their findings clearly draw connections between the perceptual placement of sound and it being favoured (in some cultures) for caregiver-infant interactions. The reason for this is that the 'object' of the speech sound was the voice of the speaker, but this also meant that, from the perspective of the listening infant, the sound was located at a distance, the impression of which encouraged greater autonomy in the infant. At this point, however, I do feel compelled to make a quick digression to observe that this does not equate to scientific validation for not hugging your children. I'm happy to go on record as saying people should hug their children.

Asking where sound is can feel like an unusual question to ask, but it is one that to this day sparks fierce debate in undergraduate lectures. It speaks to one of the most fundamental illusions of auditory perception, namely the experience of sound at its source. This illusion should be immediately available to most readers, simply by attending to any sound (ideally, a continuous sound) in your environment. If you focus your attention on that sound, you will almost certainly feel as though it is located at a specific point, away from you. Now try to disconnect that same sound from the point in space where you have located it to. It should be difficult, if not impossible. This is problematic, however, when we consider the natural properties of soundwaves and the mechanics of hearing. Our ear is not at the exact same point in space as the source and the soundwave has had to travel from the point of origin to the point of our ear for us to hear it.

More recently updated in 2020, Casati and colleagues' entry *Sounds*, in the Stanford Encyclopaedia of Philosophy (2005), is a useful primer for understanding the various competing theories of sound. The entry presents a surprisingly large number of perspectives based on the question *"where is sound?"* grouped as either proximal, distal, medial, or aspatial. The term proximal relates to the idea of close proximity. It describes theories that suggest sound is located at the listener. Within proximal theories, researchers have further debated over whether sounds are best thought of as sensations (sound exists at the point at which we hear it) or as proximal stimuli (sound exists *around* the point at which we hear it). The fine print of these theories can get very complex very quickly, with many relying upon hypothesised real-world scenarios that other researchers have subsequently countered, unpicked, and sometimes debunked; making the process of advancing the theory of sound feel like less of a stroll down a straight path and more like being lost in a hedge maze. The take-home of sound-as-sensation theory is that, unlike the acoustic definition, it accounts for times when we hear a sound in the total absence of a soundwave. The buzzing of tinnitus is a well-known example of this that you may well have experienced at some point in your life, though hopefully not for too long. In this instance, the tinnitus ringing *is* sound because it can be felt as a distinct sensation.

Medial theories bring us back to the acoustic definition. They assert that sound is the medium, or the space within which both the source and the listener are present. Within the medial theory, some assert that sound is best understood to be a mediated event, caused by the actions of an object or the interactions of multiple objects. These events exist within the space but also affect it, making it resonate, to effectively become a property of the space itself that is then received by the listener. The alternative to mediated event theory is wave theory (aka, the acoustic definition; aka, sound as a soundwave; aka, the theory with more names than a laconic tv villain) which we have reviewed the principles of already. Wave theory relates to event theory by acknowledging the relationship between source object(s) and space but extends it by designating the resonating property of the space (the soundwave) as an object in its own right. We already mentioned the issue of tinnitus (I also recommend looking up 'exploding head syndrome[2]) as a limitation, but Casati and colleagues' review outlines several further limitations, including a lack of explanation for circumstances in which: (1) two soundwaves have different physical properties but sound perceptually identical; (2) the frequency exceeds our 20–20,000 Hz hearing range, yet we still experience an auditory sensation; or (3) one soundwave can feel very different when heard on multiple separate occasions or by two different people, even if they are receiving the same soundwave at the same point in time.

Next is distal theory. Where medial theory brought us to the acoustic definition, distal theory returns us to that point where you may remember being instructed to hug your children. It places sound at or around the source. Within distal theory, perspectives jostling for dominance include (but are not limited to) property theory, located event theory, and dispositional theory. Again, the fine details differentiating each of these perspectives can be a little inaccessible so we shall stick to the broad distinctions between them. Property theory is strikingly similar to event theory, only the *property* in this case belongs to the object and not the resonating space. When our dog barks in the park, according to property theory, the barking sound is a property of the dog. In contrast, event theory would assert that the barking sound is a property of the space within the park, caused by the dog. Another distal perspective, located event theory, suggests that sound is an event, but one happening *to* a material object, or objects. This theory is best to think of as a strong advocate of the answer to that Guardian poll about trees falling in woods being an emphatic 'yes'. It asserts that the absence of an environment that enables soundwaves to propagate and reach a listener does not mean that an object is not causing sound, it is simply a sound that cannot be heard. A good, if improbable hypothetical to consider this is to imagine a vacuum between us and our barking dog. The dog is in one resonating space, and we are in another, but the soundwaves cannot physically pass between the spaces. In this instance, would you say that there is no sound, or that there is indeed sound but that we simply cannot hear it? Located event theory would argue the latter.

The last distal theory of sound we'll cover here is known as the dispositional view and is a personal favourite – not necessarily because I agree with it over competing theories, but because of its use of a wonderful sound-related word: *thwack*. As Casati and colleagues (2005) no doubt enjoyed writing: "A good thwack (an impulse that contains all relevant frequencies) is considered to be to sounds what white light is to colors". The principle of the dispositional view is that objects possess sound irrespective of whether they are presently vibrating (due to receiving a good thwacking) or not. From this perspective, I possess sound even if I do not speak, much in the same way that a green apple in a completely darkened room could still be argued to be green.

So far, we have reviewed the details of several competing theories across three different answers to the question *where is sound?* The possible answers being: sound is at the object (distal), sound is between the object and listener (medial), and sound is at the listener (proximal). This seemingly covers all the basics but leaves a space for one more possible answer, that sound is not actually anywhere. This is the aspatial theory of sound, the fundamental argument of which is that sounds are not intrinsically spatial, or at the very least, our perception of them is not. One key point underpinning aspatial sound theory is the issue of position being relative. The sound of the barking dog is to my left, but only because I am to its right. The sound is close by, but only because I am close to it. Without my position being what it is, the sound cannot inherently possess spatial information. Pasnau (1999) describes a scenario in which a person sitting indoors hears the sound of a bird chirping and, although unable to pinpoint the bird's precise location, can easily hear that the bird is outside. Pasnau's explanation is that it is precisely because the sound *is* outside that the person hears it as such, but there is a problem with this explanation. How do we know the person hears the bird to be outside based on the purely objective acoustic properties of the sound and not due to the person's innate knowledge and expectation that wild birds do not typically reside indoors? Did the person actually *hear* a bird chirping outside, or did they simply hear a bird chirping and perceptually add the localisation in the same moment?

The academic discussions on the nature of sound, both what it is and where it is, are yet to reach an agreement and much more than what is written here has been considered. Strengths and weaknesses, hypotheticals, and counter-hypotheticals have been discussed, as researchers seek to uncover the most accurate explanation for the phenomenon of sound. At this point, I do hope you were not expecting a definitive answer in this book, because there isn't one. However, returning to the purpose of the book, comprehensively explaining sound is not the goal. Instead, the intention is to look for differences in perspectives across different cultures. Do European cultures favour a distal, instantiation view of sound whilst Asian cultures relate more to the idea of sound as proximal stimuli? What is it about their culture, or *habitus*, that they believe affects this preference? Does this preference impact how they derive meaning from sound, how they feel it, or how they express it? Are there any practical applications or lifestyle benefits to having a particular perspective on the *what* or the *where* of sound?

Extending sound theory even further

You may or you may not feel that the above review of sound theory was a concise one. If not, I'm afraid we're not done. Whilst these philosophically and physically oriented perspectives provide substantial insight and platforms for debate, they are arguably very focussed, attempting to explain sound in relatively specific and finite terms. However, many researchers, particularly within the discipline of sound studies, have struggled to accept any single theory on sound, finding certain aspects within each that they agree with and other aspects that they do not. Their conclusions have become a broad contention that the truth comprises numerous elements from several theories, but also extends beyond them to incorporate further features.

Beyond the acoustic definition

In Moore's leading psychoacoustics primer *An Introduction to the Psychology of Hearing* (2012), a significant chunk of the page count is devoted to comprehensively explaining the mechanics of soundwave propagation and reception, and rightly so. Whilst there may be disagreement between those who equate sound with a soundwave and those that do not, very few would argue that such mechanics are not at least part of the overall equation. But if there are other parts to this equation, precisely what are they? Moore structures his review to compartmentalise sound into seven discrete elements: frequency, loudness, temporality, pitch, space, pattern and object, and speech. The narrative drawn by Moore is one that moves us as readers from the physical environment, to externals of the human body, to its more internal mechanics, followed by the processes of the nervous system and the brain, before arriving at the non-physiological matters of the mind.

The acoustic definition has certainly afforded us significant benefit, both historically and in continuing contemporary sound research and practice. However, and as you might expect, once modern philosophy turned its attention to the subject, limitations in the standard view began to emerge. In 1999, Pasnau argued that the standard view is built upon two contradictory assumptions: (1) sound is the object of hearing and (2) sound is a quality of the surrounding environment, not the object that makes the sound. His argument for this contradiction is that in accepting the acoustic definition, we are essentially saying that sound both is and is not the object that creates it. In his essay, Pasnau accepts the first assumption of the standard view, that sound is the object of hearing, drawing upon the historical muscle of Aristotle (c. 350 BC): "sight has colour, hearing sound, and taste flavour". What is interesting here is that what could be understood to be a fundamental absolute on the meaning of sound may, in fact, be a simple trapping of language. Effective communication within the rules of language constrains us to classify phenomena in terms of subject and object. But what if doing so could, in some instances, obscure the truth of something rather than reveal it? The problem here is with the word 'object'. Various perspectives on sound do not unpick the meaning, or meanings, of a sound object. They stick close to Aristotle and argue sound to be the object of hearing but then seem to convolute that statement with the source object.

As Goldsmith (2015) points out, nobody should suggest that our experience of sound is not influenced by the complex interplay of fundamental amplitude and frequency waveform parameters, but rather that it would be wrong of us to try and reduce our experience to them. To do so would be to dismiss various additional elements that contribute to the experiential nature of sound. This does present us with a general agreement that the principles of the acoustic definition are largely correct and central to the phenomenon of sound, but they are not the complete story. Continuing down the rabbit hole, let's look at some of the key theories that extend the acoustic definition and seek to provide a more comprehensive understanding of sound. We start with a closer inspection of a pivotal component – the listener.

What is listening and how is sound controlled?

Is there a difference between hearing and listening? If so, what precisely is the nature of that difference? If hearing is the fundamental physiological mechanism by which we receive and transduce soundwave stimuli, then how is listening different? In the context of dealing with sound stimuli in particular, academic literature broadly associates listening with control and examines it within that boundary. However, it is worth noting that other writing on listening approaches the concept in much the same way. As observed in a paper by Crawford (2009), listening has broadly become a metaphor for paying attention. It has less to do with sound per se and more to do with effective communication, well researched in contexts such as political leveraging of the phrase "I'm listening". This brings up matters such as willingness to consider, tolerate, accept, and respond to opposing viewpoints. Equally, it connects the act of listening to an ability to control a sound scenario, be it directing our attention towards a particular source of sound, being analytical or critical of a sound, or maybe a willingness to be quiet and instead allow someone else to make sound. This brings us to our third key question, one that considers the issue of control in our relationship with sound.

To what extent do we have control over our sound environment? Do some of us exert more control than others? Do other aspects of the habitus influence our potential, and our desire, to have greater or lesser control in our relationship with sound? How much cross-cultural consistency is there when it comes to a person's attitudes towards listening and to controlling sound? Sterne's foundational text *The Audible Past* (2003) draws a clear distinction between hearing and listening. This distinction is grounded in the notion of listening being more akin to a skill, described by Sterne as 'audile technique'. Deployment of audile technique is largely explored in professional contexts such as science, medicine, and engineering. As any technique, it can be refined through practice, to the extent that highly capable individuals may be described as 'virtuoso' in their audile ability. It is also characterised by the intentional separation of auditory information from other incoming sensory cues and being for the general purpose of representing an acoustic space, specifically so that space can be reconstructed to reveal a positive physical change to some aspect of (or object within) the environment.

There are many examples that could help illustrate audile technique, but one that features in recent memory concerns me and my car mechanic. It was a cold and wet January morning. After noticing repeated losses of power whilst driving, my car found itself hoisted on a hydraulic jack with its engine running. The mechanic and I stand underneath the vehicle, both of us peering upwards. The soundscape is a disorientating cacophony of what to me is noise. The whir of the hydraulic jack continues, a radio blares, and another mechanic is hammering away at some other poor soul's car. Outside, a jackhammer is relentlessly decimating pavement tiles and my car's engine is merrily reverberating, the sound bouncing repeatedly off the numerous concrete surfaces of the garage. The mechanic continues to peer at the underside of my car's engine. For a moment, his brow furrows, before he takes a wrench to some hitherto unknown component. The engine noise changes in some way I cannot describe. He turns to me and pronounces "there you go". For me, this act of what might as well have been sorcery most certainly qualifies the title of virtuoso. The mechanic was seemingly able to perceptually shut out all extraneous sources within the soundscape to focus on the engine. He was then able to analyse the attended sound to diagnose a problem and carry out a solution. Finally, he was able to determine if the soundscape had been altered based on his intervention, and if that alteration was indicative of the problem being solved.

Typologies of listening

In *Sonic Skills* (2019), Bijsterveld considers listening in its professional usage within a factorial typology based on purpose and approach. Bijsterveld identifies three listening purposes: monitoring, diagnostics, and exploration. Monitoring works from an initial assumption that all is well, to be confirmed if there are no deviations from a 'healthy' soundscape. Diagnostics presumes a problem but does not yet have confirmation or knowledge concerning its precise nature (my mechanic). Lastly, exploration involves neither positive nor negative presumptions but is instead concerned with discovery. Bijsterveld's listening approaches comprise synthetic (general acoustic impressions of a soundscape or individual sound over time), analytic (specific acoustic properties or features of an individual sound or soundscape at a limited point in time), and interactive (acoustic changes directly attributable to an intended action by the listener). Audile technique is a great example of connection between the habitus and sound experience. In this case, it raises important questions regarding how our professional practice and training may be influencing out relationship with sound.

Sound studies has also directed its interest towards the subject of our everyday lives. One example of this concerns the architecture and the pageantry of sound environments. In 2008, Rebelo and colleagues published a *Typology for Listening in Place*. Their work argued that the nature of listening could be categorised into three groups, based on key features of the architectural space and the culturally established expectations that the space proliferated. The first is the 'theatre of listening', a term applied to scenarios in which the projection of the sound is both known and constant. An actual theatre provides the most obvious example. Here, the location of the sound source and its projection are established in advance by way of the architecture. The sound objects related to speech will all be positioned upon the stage and the speech itself is largely directed outwards, towards us as the audience. Those speech-sound 'objects' are the actors upon the stage, a clearly designated role with an established relationship to our own role as the spectators. The music-sound objects we will knowingly find in the orchestra pit. Although our interpretation and subjective critique of the sound in a theatrical performance may differ, the physical characteristic of the experience is objectively homogenous in that it is broadly unchanged between each listener and between repeat performances. This does not mean that the experience is non-interactive or purely passive. All sound experiences can be argued to be both active and interactive, but the devil is, as always, in the detail. In the theatre of listening, we cannot easily physically interact with the sound, save for putting our hands over our ears or selecting our seats carefully, and our active listening is therefore largely affective and cognitive.

In contrast to the theatre of listening, the 'museum of listening' presents a more fragmented sense of sound object location and soundwave projection, as the listener is free to move around the space as they wish. They may attune to specific sounds in a variety of orders and spend more time listening to some sounds than to others. They may even choose to avoid certain sounds altogether. This facilitates some physical heterogeneity of experience, but the boundaries and architectural characteristics of the space are similar to the theatre in that they are also known and controllable. The listener may be able to move about with freedom and autonomy, but the nature of the sound space is largely static. Sources (exhibits) are not erected, pulled down, or switched out whilst the listener is present, meaning the listener can (in most cases) reliably return to a previously visited point and experience the same sound.

Finally, the 'city of listening' moves us to the fully fragmented end of Rebelo and colleagues' listening spectrum. The boundaries that define the outline of a city can be obscure and debatable. The sound objects within the city may include some reliable entities, but most are dynamic, moving about the city or in and out of it completely. A city of listening experience is defined by having the most uncontrolled characteristics and the greatest opportunity for listeners to physically interact with the soundscape.

Whilst the theatre, museum, and city are certainly archetypal examples of these forms of listening, they are not the limits of such examples. A theatre listening experience may, for instance, describe a cinema, a lecture, a personal audio player, or even a football match. Museum listening may denote a typical video game or virtual reality experience (procedurally generated content possibly moving us more into 'city' territory), whilst a city listening experience could be pushed to further extremes of unknowability, considering environments such as an active warzone or the site of an unfolding natural disaster. The key point to raise here is that such substantial differences in our listening experience are resultant from the architectural and organisational aspects of our local environment, what we shall later refer to as 'place'. Precisely what effect could walking through a dynamic and ever-changing city each day have upon our relationship with sound, compared to if we lived and worked in a quaint countryside hamlet? Do some of us prefer theatre listening to museum listening scenarios? What cultural factors might affect this? Do we adapt to certain listening scenarios with repeated exposure, and does this change our personal preferences towards sound? How might our attitudes, expectations, and judgements on sound differ if we predominantly control our sound spaces, using personal music devices with noise cancellation when moving between familiar soundscapes, as opposed to embracing sonic chaos?

The function of sound

Let's return to our imaginary park, and our equally imaginary dog (who is still, imaginarily, barking), but now a tweeting bird has joined in. If I were to ask you to tell me what you can hear whilst we are sitting in the park, you may respond with 'a bird' or 'a dog'. Of course, in this scenario, the task is explicitly to identify, which encourages you to direct your attention and classify your auditory scene in terms of sound objects. Now say I asked you to describe in more detail the quality of the sounds you can hear. In this instance, identification is no longer the main requirement of the task, and you must now consider the sounds over time to assess and express their qualities. In this context, understanding the sounds merely as objects would seem very limited, and your response will almost certainly be richer in information. This is a particularly important point to consider, both in terms of its application to raising awareness of different meanings and expressions of sound between cultures, and also in terms of understanding sound in general. If we agree that listening requires attention, then theories of listening function assert that our purpose for listening will direct our attention within a soundscape and that this will impact not only what sounds we listen to, but also why we listen to them. This brings us to a further question. *Why* is sound? How do we define our relationship with sound in terms of function?

The notion of listening function overlaps heavily with 'modes of listening'. The basic principle of listening modes is that we can attune to sounds in various ways, or modes. Each mode provides us with a different type of information. Two fellows arguably bearing substantial responsibility for our current understanding of modes of listening are Tuuri and Eerola (2012). They present us with nine individual modes of listening, which they organise into a taxonomy based on the degree of conscious attention and consideration we pay to any given sound. The modes are structured into three groups: experiential (pre-attentive/subconscious and automated responses immediately following the commencement of a sound), denotative (quick extractions of meaning), and reflective (slower, more considered, and analytical extractions of meaning). Within the experiential group, the most immediate listening mode is an immediate bodily response, such as muscular tension or a startle reflex in response to a sudden sound. This is followed by a kinaesthetic response that provides the listener with an immediate sense of a sound object's position and motion. Lastly, connotative response extracts free-form associations (i.e., the first thing the mind thinks of) following the commencement of a sound event. Within denotative modes, the details we can extract from the sound include identification of the source, any indication of instruction to which we might need to immediately respond, empathetic information that could infer the emotional state of the source, and any sense of the sound or its source's purpose or basic function. The reflective modes are analytical appraisals, which can often continue long after the physical sound has ceased. They included reduced analysis, which considers sonic characteristics (loudness, periodicity, pitch, timbre, consistency, consonance/dissonance, etc.), and critical analysis, which describes moments when we apply a value judgement to the quality or appropriateness of a sound in context.

The above modes of listening could, at least in theory if not practice, all be applied to a singular listening experience in which the listener immediately reacts to a sound but then considers it to increasingly analytical degrees. Continuing ever so relentlessly with our hypothetical park and barking dog scene – here, our listener was not anticipating hearing the barking and their body, within a split second, muscles across their body contract, priming them for action. Their eyes stretch wide open, and they draw a short and sharp breath. The sound appears to be directly in front of them and moving closer. Our listener responds without thinking, contorting their body backwards and away from the source. As they do so, the most immediate thought in their mind is *danger*. Within the next few moments, our listener considers several denotative features in no specific order. They identify the source as a dog, the instructional meaning as "I want to play!", the emotion as playful, and the purpose as "that's just what dogs do when they're happy!" These intermediate modes, in this instance, all provide reassurance. Our listener's sense of impending doom subsides, and their heightened emotions abated. As the dog continues to bark, they observe the loudness and deep pitch of the bark. They then note the diminutive size of the dog and are impressed that such a small animal could make such a big noise.

It is important to note that the scene depicted above is formed of various arbitrarily chosen contextual factors, any of which we could change the value of to create a different listening scenario even though the acoustic properties of the soundwave could remain completely unchanged. For instance, let's say our listener could also see the dog in front of them and is watching the animal closely. Moments before the dog begins to bark, our listener observes the change in its posture as the dog effectively telegraphs its impending exclamation. The acoustic properties of the bark remain the same and yet our listener's body does not tense up. They don't suddenly move in any direction, nor do they experience any pre-attentive sense of danger that may evoke a fight-or-flight response. Alternatively, say our listener had somehow never been introduced to the idea of a dog. They quite simply could not draw a causal connection between the sound and a source object of which they have no concept. Now it is just a 'thing' that is making a noise. As a final example, say our listener had very little experience of dogs, to the extent that they had no ability to differentiate the various denotative features of the barking from the acoustic properties of the sound. To them, all barks sound the same and so as an instinctive precaution, they interpret any barking to denote an angry emotional state, the purpose of providing a warning, and the instruction "get back, or I'll bite". Our listener makes a run for it. Our playful dog emits a disappointed whimper, turns tail, and goes hunting for a new friend.

As noted previously, modes of listening can also be thought of as sound functions. This is because each mode clearly provides a discrete value to the listener, the specific nature of these values being heavily determined by context. In reflexive listening, the sound functions as an immediate warning system, forcing the body to attend to the sound both physiologically and cognitively in case the sound is indicative of a threat. In kinaesthetic listening, the sound functions as a localisation tool by providing near-instant locational and motion information relative to the position and orientation of the listener; arguably the most vital second stage in threat assessment after the potential presence of a threat has been identified. Likewise, in connotative listening, the function of the sound is to provide any further information that might aid the listener in responding to the threat, but specifically information that can be extracted and interpreted almost immediately. Across the denotative modes of listening, causal listening presents the function of identifying objects in our local space, whilst functional listening provides understanding of events, how objects are behaving and interacting. Semantic listening enables us to receive and respond to instruction, and empathetic listening allows us to nuance our response behaviour based on our emotional awareness of a situation. Finally, reduced and critical listening enable us to assess sound's acoustic properties with relevance to other known information, as a means of understanding more about a source, situation, or local environment, then using that understanding to form judgements.

The bulk of research concerning how we listen is, presently, either largely theoretical or focussed upon musical listening. For example, a 2010 study by Brattico and colleagues provided empirical evidence of a significant consistency between listeners in terms of identifying the correctness and describing the congruousness of a chord within a sequence. Their results also revealed consistency in reaction time between responses to liked and disliked chords, with the average response time significantly shorter following a disliked chord. Using electroencephalographic data, Brattico and colleagues were also able to show clear and consistent differences in objectively measured neural activity when participants were asked to judge the musical content in affective (like/dislike) or cognitive (correct/incorrect) contexts. Here it was shown that the auditory processing pathway within the brain will, quite literally, light up differently when listening to the same sound if we are asked to evaluate it in different ways. Such studies indicate that whilst a huge range of factors may influence our experience of (and interactions with) sound, we are all human, and there will always be certain universals in our relationship with sound.

Function provides us with an explicit framework for understanding how a person may interact with sound in terms of their physiology, their emotions, and their cognition. Functions more closely relevant to our physiology and biology may be less likely to differ greatly between cultures, though even that is not beyond the realms of possibility. Not all sudden noises within a controlled environment will have the same experiential effect (making us jump, raising our heart rate, etc.) on multiple listeners. The potential degree of difference only becomes larger in terms of the denotative effects of a sudden noise (such as our emotional interpretation of the sound source), and even more so when we look at reflective effects (a critical judgement on the extent to which the sound was scary, for example). This helps us to start formulating some sensible hypotheses for comparing interaction with sound across cultures. It also presents us with some opportunities to be surprised, a key ambition of this book.

Acoustic ecology

Earlier in this chapter, we looked at how the environment was an important feature of auditory perception, briefly referencing the notion of 'place' and its possible effects on our relationship with sound. This issue resonates keenly with acoustic ecology. Building upon key arguments such as Truax's (1984) assertion that sound mediates our relationship with the environment, Wrightson (2000) explains how acoustic ecology relates heavily to matters of soundscape. We shall return to soundscape in Chapter 5 but, for now, the salient point is that how our own actions and the changing state of the environment (both that which results from our actions and that which does not) all influence our sonic world. This, in turn, affects both us and our environment. Wrightson's 'noise generator' exemplifies this, describing an escalating cycle in which soundscape noise increases, forcing us to increase our own sound loudness just to be heard, which in turn forces the soundscape noise to increase even further, ad infinitum.

Later research has expanded upon acoustic ecology models with the introduction of 'virtual sound'. This can be better explained by first introducing the notion of diegesis. The concept of diegesis has a long history, dating back to Plato's *Republic* where it was used to describe the relationship between the fictive world of a narrative and the real world. The term 'real world' is itself a little problematic as diegesis can be used to describe the real world within a narrative, but in that moment, the description of the real world is not the real world itself. Other ways of phrasing the real world include 'actual' and 'physical', as different writers grapple with this metaphysical confusion to try and understand precisely what the meaning of diegesis within our existence is. By coincidence, it is sound that provides us with a powerful means of depicting diegesis. Imagine yourself in the theatre, indulging in a stage musical. The actors upon the stage produce sound, their speech, and actions audible to you and to the rest of the audience. At the same moment, the orchestra is in their pit, producing their own sound, an incidental score accompanying the action, reciprocating its tone, affect, and pacing. The sound upon the state is diegetic, that is, it exists in the fictional context of the narrative. The actors can respond to each other's sound because it is present in the narrative world alongside the characters they are portraying. The sound from the orchestra pit, however, is non-diegetic. It does not exist in the fictional context of the narrative. It exists with the audience, in the real world. You wouldn't expect an actor upon the stage, during a performance, to comment that the second-chair violin sounded a little flat in the fifth bar, because that violin is not present in the narrative world. But then, a character upon the stage begins humming a melody heard moments ago upon a violin from within the orchestra. Now, where does the orchestra belong? Is it of the 'real' world with the audience, or of the fictive world with the narrative and characters? Did its diegesis change? What is happening?

Diegesis has found common usage in both historical and modern discourse, starting with literature and theatre before being considered with regard to cinema (see 'fourth-wall breaks') and digital games, the latter of which is the medium that takes us from the traditional models of acoustic ecology to a model of *virtual* acoustic ecology (VAE). Coined by Grimshaw (2007), VAE considers ways in which our relationship with the environment, mediated by sound, is changed when you add the diegetic dimension of virtual worlds (which still of course exist within the physical/actual world) into the mix. One of Grimshaw's key observations is that VAE offers new ways of experiencing and interacting with sound. It also presents new listening functions. One of these functions, navigational listening, is essentially the use of sound cues to localise oneself and to navigate around a space. Here we have an extension of the kinaesthetic and localisation functions discussed a few pages ago, with clear relevance to our everyday interactions with sound in the real world. We utilise navigational listening when we ask a friend to call out to us so we can get a sense of their position and move ourselves closer to them, or when we assess the sounds of a car's engine to direct our angle and rate of motion as we cross the street. However, it is worth pointing out that no formal identification of the navigational function of sound was identified until Grimshaw considered how individuals utilise sound in a virtual context.

Grimshaw's VAE adds further specifics to navigational listening, asserting that individually attended sounds (or characteristics of a sound) can fulfil individual navigational functions, namely attractors, retainers, and connectors. When our friend calls out to us, their call is an attractor, directing us towards the source. When we cross the street, the car's engine, based on its localisation and kinaesthetic characteristics, may provide the function of an attractor, encouraging us towards the pavement more quickly as we interpret inactivity to lead us to being possibly hit by the moving vehicle. Should the sound indicate a likely collision if we continue to cross, the sound will hopefully act as a retainer, halting our movement and keeping us in our current location until the vehicle has passed. The connector function is more temporal. It describes circumstances in which a sound is providing ongoing feedback in relation to our continuing movement. It tells us that we are transitioning from one point to another. Returning to our scenario of crossing the street, this would describe the changing positional characteristics of the engine sound as the listener continues to cross. Such characteristics could include the balance between the left and right ear and the overall perceived loudness of the vehicle, and we may also use our experience of pitch derived from engine revs to give an auditory impression of the vehicle's speed. This ongoing stream of feedback can allow the listener to nuance their movement to help them cross safely. This scenario has in recent years prompted several studies seeking to investigate the potentially enhanced risk of injury to pedestrians wearing noise-cancelling headphones (Wachnicka & Kulesza 2020; Mikusova et al. 2021). The above mechanism certainly points to this being a valid hypothesis, but, at the time of writing, empirical studies appear to be inconclusive, though it remains a very sensible premise that you are no safer crossing the road with your fingers in your ears than you are putting your hands over your eyes.

In addition to navigational listening, VAE also extends beyond the fundamentals of getting from A to B, with sound functions that identify specific matters of space and time through sound. So far, we have looked at how modes of listening can provide several strands of information on a source object, but it is also possible to gain even further understanding of our environment through sound. As defined within Grimshaw's (2007) model, a *choraplast* function emerges from the interaction between the soundwave and the acoustic properties of the 'resonating space' to give information on that space. If we perceive a sound to have a persistent reverberation with imperceptible delay, this can be indicative of a small, enclosed space comprised of reflective surfaces and hard materials. This is how people can sometimes catch you out when you're speaking to them from your bathroom, even though no taps have been run nor has a toilet been flushed. Alternatively, a persistent echo with a long delay suggests a large open space, also comprising hard surface materials – the 'White Cliffs of Dover' effect. A choraplast can therefore provide not only a nominal indication of your environment (a church, studio, field, etc.) but also information concerning its physical properties such as scale and material composition.

Where the above choraplast function is descriptive of space, *chronoplast* and *aionoplast* functions are descriptive of time. A chronoplast refers to sound that indicates more discrete and immediate temporal issues within the environment to which the listener needs to respond. It may also encourage them to control their rate of motion. Alarms and ticking clocks are archetypal examples, but other common everyday examples include the series of beeps at a pedestrian crossing that provide a sense of time remaining to safely cross, and the fundamental effect of musical tempo on our rate of motion, be it dancing along or merely tapping your foot. Whilst the aionoplast function of sound also deals with matters of time, it is not concerned with immediate temporal issues but longer periods or epochs. The sounds of transportation are a good example of this and return us to the everyday scenario of crossing the street. In this example, the attractor, retainer, connector, and chronoplast functions of sound would work in broadly the same way but, crucially, would be modulated by the aionoplast function. As of 2021, many streets feature a combination of electric and combustion engine vehicles. We only need to travel back a few years, and the soundscape in this context would be almost entirely combustion engines, whilst if we travel hypothetically forwards by roughly the same number of years, it is likely that sound scape will be almost entirely electric. Aionoplast function can be exemplified most potently when considering the soundscapes of alternate historical periods. Consider, for example, the aionoplast difference in the soundscape of our street, were it in the Middle Ages. The acoustic interplay between rubber and tarmac would be replaced by metal horseshoes upon stone-cobbled streets. Animal vocalisations would flood the soundscape, comprising the horses carrying out the heavy lifting but also the array of pigs, geese, hens, and more being transported to market. Of course, many present-day locales around the world protect the historical features of their environment, from the cobbled streets to the thatched roofs, raising the potential for experiencing something of a sonic time warp when travelling between different places.

The above elements conceptualise sound in ecological terms and provide us with several further things to look out for during the interviews. Matters of space and material have clear implications for questions concerning our relationship with sound within the home, in the workplace, and out in the world. How those matters influence our movements within those spaces is also an important question to explore whilst temporal elements encourage consideration of our structure and routines; and even how long-term changing soundscapes may be subtly influencing lasting changes in our daily lives.

Chapter 1 summary: questions begetting questions

The objective of this opening chapter was to review a range of prominent theories relevant to sound, the aim of which was the realisation of an initial list of questions that could be proposed to our contributors. The following questions were formulated based on the theories and findings that we have discussed throughout this chapter:

- What does sound mean to you?
- What does listening mean to you?
- How would you describe the function of sound in your home, work, or out in the world?
- How do you control sound, both at home and out in the world?
- To what extent do you feel your professional practice influences your relationship with sound?
- Can you describe your local 'place' in terms of sound, and do you think your local environment influences your relationship with sound?

The above is not the complete set, and we have yet to cover all the research that generated numerous further questions on matters of culture, communication, space and place, or professional practice. However, considering the volume of theory I've already thrown at you, it felt almost cruel to subject you to the entire review in one sitting and so, the remainder of the literature we shall return to in subsequent chapters that examine the above topics individually and help contextualise contributor responses.

It is fair to say that there is *a lot* of theory on sound. Numerous competing perspectives and definitions jostle for position, some seemingly straightforward and matching commonplace understanding, others bordering on mind-bending. But are our responses to these perspectives unique to us? To what extent has our personal history, current place in the world, or any other factor for that matter, determined our reaction to the theories? Do we all fundamentally understand the meaning of sound differently? Whilst the questions listed on the previous page were to be posed to our contributors, it is this latter set of questions, combined with the above that we hope will ultimately reveal new insights into the human relationship with sound.

Notes

1 https://www.theguardian.com/notesandqueries/query/0,-82446,00.html (accessed 18.03.2021).
2 Type "exploding head syndrome" in your search engine. It's supposed to be rare but, honestly, everyone I describe it to tells me they've experienced it in some form.

References

Bijsterveld, K. (2019). *Sonic Skills: Listening for Knowledge in Science, Medicine and Engineering (1920s–Present)* (p. 174). New York: Springer Nature.

Brattico, E., Jacobsen, T., De Baene, W., Glerean, E., & Tervaniemi, M. (2010). Cognitive vs. Affective Listening Modes and Judgments of Music–An ERP Study. *Biological Psychology*, 85(3), 393–409.

Casati, R., Dokic, J. & Di Bona, E. (2005). Sounds. In E. N. Zalta (Ed.) *The Stanford Encyclopedia of Philosophy*. https://plato.stanford.edu/archives/win2020/entries/sounds/ (accessed 18.05.2021).

Crawford, K. (2009). Following You: Disciplines of Listening in Social Media. *Continuum*, 23(4), 525–535.

Di Bona, E. (2019). Why Space Matters to an Understanding of Sound. In T. Cheng, O. Deroy, & C. Spence (Eds.) *Spatial Senses: Philosophy of Perception in an Age of Science* (pp. 107–124). New York: Routledge.

Freigang, C., Richter, N., Rübsamen, R., & Ludwig, A. A. (2015). Age-Related Changes in Sound Localisation Ability. *Cell and Tissue Research*, 361(1), 371–386.

Goldsmith, M. (2015). *Sound: A Very Short Introduction* (Vol. 451). Oxford: Oxford University Press.

Grimshaw, M. N. (2007). The acoustic ecology of the first-person shooter (Doctoral dissertation, The University of Waikato).

Hammersley, M., & Atkinson, P. (2007). *Ethnography: Principles in Practice*. New York: Routledge.

Kärtner, J., Keller, H., & Yovsi, R. D. (2010). Mother–Infant Interaction During the First 3 Months: The Emergence of Culture-Specific Contingency Patterns. *Child Development*, 81(2), 540–554.

Merton, R. K. (1959). Introduction: Notes on Problem-Finding in Sociology. *Sociology Today*, 1, 17–42.

Mikusova, M., Wachnicka, J., & Zukowska, J. (2021). Research on the Use of Mobile Devices and Headphones on Pedestrian Crossings—Pilot Case Study from Slovakia. *Safety*, 7(1), 17.

Moore, B. C. (2012). *An Introduction to the Psychology of Hearing*. Bingley: Brill.

Pasnau, R. (1999). What Is Sound?. *The Philosophical Quarterly*, 49(196), 309–324.

Rebelo, P., Green, M., & Hollerweger, F. (2008, May). A Typology for Listening in Place. In *5th International Mobile Music Workshop. 13-15th of May* (pp. 15–18), Vienna, Austria.

Sterne, J. (2003). *The Audible Past: Cultural Origins of Sound Reproduction*. Durham: Duke University Press.

Truax, B. (1984). *Acoustic Communication*. Norwood: Ablex Publishing.

Tuuri, K., & Eerola, T. (2012). Formulating a Revised Taxonomy for Modes of Listening. *Journal of New Music Research*, 41(2), 137–152.

Wachnicka, J., & Kulesza, K. (2020). Does the Use of Cell Phones and Headphones at the Signalised Pedestrian Crossings Increase the Risk of Accident? *Baltic Journal of Road and Bridge Engineering*, 15, 96–108.

Wrightson, K. (2000). An Introduction to Acoustic Ecology. *Soundscape: The Journal of Acoustic Ecology*, 1(1), 10–13.

2 Sound and research

Albert Road, Southsea. 14th of August 2021. Early evening

After crossing the road and continuing down the high street, I reach what is a bit of a night-life hub at the lower end of Albert Road. It's a warm evening and the venues have spilled out their seating into the street, with many people talking and drinking outside. The soundscape is filled with a million tiny clinks, scuffles, tings, and scrapes that form a singular mess of sound, with the voices easily the standout feature. And yet, the sound remains very restrained, conversational, and emotionally neutral. Chalk up a point for British civility in a public space. Then I consider the time. It is pretty early in the evening. Give it an hour, two at most. It will not sound so restrained and neutral then I'll wager. I fear I may have visibly chuckled smugly to myself, before the dawning realisation that my mind is clearly overcompensating for my jaded, middle-aged attitude. An attitude that is probably the reason I don't get invited out to places like this anymore.

The madness, the methodology, and the method

When it comes to research, few would argue with the suggestion that things can get very complicated and very confusing, very quickly. Just say the word 'methodology' to most PhD students and they'll reliably contract inwards, avert your gaze, and let out a fearful hiss, not unlike a vampire being presented with a cross dipped in garlic butter. Working in academia does, in most cases, support (/demand) increased confidence in your methodological process. However, the core nuts and bolts of designing research can still be decidedly sensitive to intellectual insecurities and feelings of imposter syndrome. Even highly successful researchers have been known to misattribute their successes and accolades to the more superficial, presentational, and performative qualities of their work – believing this masks the failures in their theory and process (Parkman 2016). Taking a moment to consider

DOI: 10.4324/9781003178705-3

methodology itself might come across as a little patronising to some, but even world-leading experts have been known to return to the fundamentals of their field from time to time, as no single aspect of scientific inquiry should be taken for granted.

I also need to make a pretty substantial confession: I have always hated qualitative research. I like numbers. I like data acquired from measures that aren't deeply susceptible to all those human frailties and biases. I don't want to talk to people. I don't want to ask them their thoughts, feelings, or impressions. I'd be much happier wiring them up to all kinds of sensors or making them perform tasks like obedient little mice in a maze whilst I collect all that wonderfully objective physiological and behavioural data with which I could then run lots of lovely statistical analyses and then simply let the resultant numbers tell me what to think. If you haven't experienced $p < .01$, you haven't lived (and $p < .05$ is for the weak). As we shall get to shortly, and although the thought pained me, the questions that were being raised from the initial review in Chapter 1 were clearly inappropriate for quantitative methods. I needed to confront my issues and ignorance of qualitative approaches. I was now on a methodological safari, starting my journey as the obnoxious idiot abroad and hopefully ending it with a better understanding of, and appreciation for, what qualitative research can achieve.

The chain of study design and its four links

Of the many research disciplines out there, each is underpinned by a particular philosophical position. Here, we are predominantly talking about epistemology, which is concerned with the theory of knowledge and methods to its construction and validation. We can also stretch our perspective wider to consider ontology; questioning the very nature of existence in relation to physical, social, and cultural contexts (Ejnavarzaia 2019). In *Ethnography* (2000), Brewer starts with such foundations, describing the philosophy of social research by way of a four-stage causal chain, with the nature of each link determining that of the next. This chain commences with the link of methodology, a theoretical framework with the fundamental function of defining link number two. This second link is a set of procedural rules that can be applied in scientific practice. Precisely how these rules are to be applied in practice is the subject of link number three, a set of research methods, techniques, tools, measures, and practices. Finally, the precise selection, configuration, and execution (or procedure) of these methods that determine knowledge generation is the fourth link that completes the chain. In effect, the third and fourth links in this chain can be thought of as a kind of 'conveyor belt' processor, into which we feed raw data acquired from the world as input, so that we can receive 'processed' knowledge as output. As researchers, our responsibility is to select the appropriate methodology, follow its procedural rules to generate the most appropriate methods, then execute those methods in the most robust way possible to ensure the most valid data and subsequent knowledge is generated.

To make things a little more complicated, the fundamental validity of a methodology and its procedural rules could potentially be (or more realistically, almost certainly are) imperfect. As Brewer points out, methodology, its construction, and its validation are under increasing scrutiny across the scientific communities. Once upon a time, criticism of a study was largely focussed on the application and execution of the procedural rules. Methodology was effectively taken for granted and rarely challenged. Nowadays we are also increasingly questioning the fundamentals.

When first thinking about research method, I had to confront the fact that I was as much of an ethnographer as I was a professional hand model. However, the central premise of the book is the human relationship with sound across the world. At least in common-sense terms, ethnography seemed to be precisely the domain in which I needed to operate. Of course, common sense and science are not always the same thing. Seeing as at least a handful of you fine readers could have an established understanding of ethnography (some may even be an actual ethnographers!), it was always going to be a good idea for me to explore the details of ethnographic methods, to confirm their relevance to the aims of the book and that I was conducting the work appropriately. And to my ethnographic expert readers: firstly, thank you for reading; secondly, I'm very sorry for each and every crime I have committed against your discipline.

The research onion

Whilst looking at the nature of study design, the four-link chain is very helpful in breaking down the fundamental, high-level components and giving us an understanding of how one leads to the next. What we now need is some indication of precisely how we are meant to move from one link to the next and how we should ultimately decide upon the best way of getting some decent answers to our research questions. In a particularly helpful work by Saunders and colleagues (2009), a high-level model of research is presented that can and has been applied across multiple disciplines. It is one of my go-to diagrams for introducing research concepts to students. It's called the research onion (Figure 2.1).

The research onion model follows but also extends upon our four-link chain of research: methodology, procedural rules, methods, and techniques. Here, we start at the outermost layer with philosophy, or epistemological and corresponding ontological foundation if you're feeling fancy. This determines your fundamental attitudes to what does and does not constitute valid knowledge. Having trust in sensory experience, collected and then analysed through a framework of logic and reason, would pop you in the positivist camp, which is also where you'll find the realists, who assert that existence is perception independent and not in the eye of the beholder. Alternatively, trust in thoughts, words, concepts, and language as valuable knowledge would sit you alongside the interpretivists. As a pragmatist, you would find yourself far less concerned with the question of whether a piece of knowledge was true, but rather if

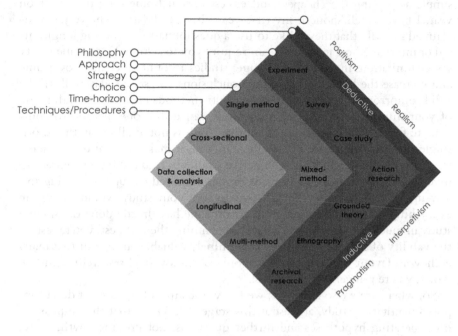

Figure 2.1 The research onion. Based on the model by Saunders et al. (2009)

were useful. The next layer of the research onion distinguished inductive reasoning, meaning the synthesis of information (observations and experiences) to formulate a truth, from deductive reasoning, which describes the use of information to prove or reject potential truths. Immediately we can see both ethnography and the ambitions of this book sitting far more in the pragmatist-interpretivist-inductive camp. We have no pre-existing truth to test yet but wish to formulate potential truths through collection and analysis of new information. This is an inductive approach. We are less interested in a singular, all-encompassing objective truth and more in the subjective experience formed from the interactions between people and sound. This puts us squarely in the interpretivism camp, but it would also be fair to label us as pragmatists. This is because, one could argue, there is little practical application to be had from amassing reams of objective, quantifiable, empirical data on sound when the whole point of the work is to better understand the nature of people – to be even more precise, people as they exist in their individual worlds of sound.

Speaking of pragmatism, the next two layers of the research onion is where practical and logistical implications begin barging their way to the front of the decision-making process. As its name implies, single-method studies require just that, whilst mixed and multi-method choices immediately raise the complexity of a study. In combination with the time–horizon layer, this can have an exponential effect on the expense of a project, with longitudinal studies often vastly longer than cross-sectional. Combine longitudinal with multi-method and you're a far braver person than I. Of course, it is not as simple as picking the cheapest and easiest option (something that, year on year, I have to tell/shout at my first-year PhD candidates, who've just proclaimed proudly that they'd like to use a questionnaire). Avoiding longitudinal or multi-method approaches can require you to acknowledge some pretty severe limitations in your work. Longer studies account for changes over time and increase the likelihood that any conclusions you state today will still be valid tomorrow. Likewise, mixed or multi-methods enable cross-validation of your information, meaning your study can utilise one method to compensate for certain limitations of another. This is not at all to say that short, single-method studies are of no value. Indeed, it makes a great deal of sense to start with smaller, more cost-effective studies to find out if it is even worth investing the much greater sums of money required to support the big follow-ups. The trick is to first discover where your study would fit within everything that has come before it. If someone has already done the smaller study in much the same way as you were planning then, at best, you're testing the validity of that previous work (a genuinely valuable endeavour that many eschew in favour of the relative glamours of 'innovative' research) whilst, at worst, you're reinventing the wheel.

So, where are we so far? Well, we know the aims of the project don't support quantitative study, at least at this stage. We know that the emphasis is on generating hypotheses and further questions, not proving anything, so, inductive reasoning. We appreciate that richness in this type of research will stem from facilitating more interpretative methods, yet we don't want the information acquired to be for its own sake. It should be able to imply some potential application, edging us a little closer towards pragmatism. That peels back the layers of philosophy and approach, bringing us to research strategy and our next port of call, ethnography.

Approaching ethnography

Ethnographic research fits neatly under the umbrella of social science. Looking to our research onion again, ethnography finds itself in the family of inductive approaches. This makes it good for generating hypotheses rather than testing them and is helpful in epistemological contexts (i.e. research questions) that value interpretive and pragmatic information. So far, so relevant, but this section unpicks a bit more of the fine print, with the aim of digging out some specific pointers on ethnographic study design, to check that we're on track to get the most valuable information for our efforts.

Qualitative v. quantitative

Ethnography is a branch of social research, specifically cultural anthropology, with roots in 19th Century Western civilisation. Anthropologists of the time felt there was a need for a sub-field of anthropology to differentiate inquiry of Western cultures from that of non-Western cultures. Study of the latter took the name of ethnography and would largely take either an overarching look upon a particular culture or would examine specific elements within it. At the time, ethnography served as a foundation for ethnology, the comparative and historical analysis of multiple cultures that endeavoured to reveal similarities and differences in human experience and behaviour across the world. As Hammersley and Atkinson (2007) point out, ethnology was steadily dropped as a discrete field of study in the early 20th century, largely due to the methods of the study becoming more integrated. Ethnographic field workers, previously responsible for conducting empirical studies and acquiring data that would then be analysed centrally by ethnologists, began conducting the more theoretical and comparative work themselves. As such, ethnography became the singular term, encompassing both ethnographic and ethnologic practices.

Ethnography was historically a qualitative field of study. However, late 20th-century authors argued the value of mixed methods and combining qualitative and quantitative approaches (Greed 1994). Subsequently, published papers and books began to present quantitative ethnography as a field of study in its own right. For some, this goes against the fundamental principles of ethnographic research. In some instances, it is even viewed as a cynical attempt to acquire wider acceptance from the scientific community by appealing to its penchant for highly generalisable conclusions at the cost of the individual richness.

As I was rather emphatic (if not, overdramatic) about earlier, I must admit to falling broadly into the quantitative camp, due to a personal preference for more objective data types and a stronger epistemological belief in positivism and realism, particularly when compared to interpretivism. I would view subjective reporting from participants as a means of reinforcing (ahem) *proper* data which, for me, was quantitative and should prioritise objectivity over subjectivity. In my own practice, I couldn't help but that colleagues seemed more impressed by the high workload associated with mixed-methods studies, as if simply putting more time and effort in was itself a good way to boost the credibility of your results. That was my most favourable view on qualitative data and, for many years, looking like I was working harder was the only motivator I had for collecting it. In my own mind, acquiring subjective reports in particular felt like a complete waste of my time. This was based on my historical belief that people are basically useless when it comes to understanding, let alone accurately and reliably describing, their own experience.

In 2010, I ran a small exploratory study that looked into the effects of different acoustic parameters of various sounds upon players' perceived intensity of a horror game experience (Garner et al. 2010). This was around the start of my second year of PhD study, and I was keen to impress so I threw a lot of methodological matter at the wall. Using a custom virtual environment made it pretty straightforward to collect basic behavioural and performative data such as completion time and the number of times the player 'ran' (virtually) rather than walked. The tests also used a variation of the *Think Aloud* protocol, which involves asking participants to verbalise their thoughts and feelings whilst concurrently performing a task or experiencing a stimulus (see Jääskeläinen 2010). My own study constrained these responses to a very simple 1 (low intensity) to 5 (high intensity) utterance whenever a prompt appeared on the screen. This was captured using a webcam that also enabled me to examine each player's body language and facial expression. Suffice to say, watching participant after participant visibly jumps out of their skin, then report a '2' for the intensity of their experience did not fill me with confidence in the validity of self-report.

Reconsidering the above experience and my attitudes towards it, I do accept firstly that allowing a single experience to sour me on a sizable chunk of scientific methodology is not exactly a hallmark of professional practice. Secondly, it is important to restate that ethnographic research, whilst capable of testing hypotheses, has a greater reputation in generating hypotheses and new questions for further study. In this context, my previous gripes lose their relevance and self-report now becomes a front runner for the purposes of this book.

The function and values of ethnography (and when to use it)

For Greed (1994), ethnography is the study of a particular 'tribe', with this term being highly flexible and facilitating inquiry at any structural level within a society. The function of ethnography is to better understand human cultures from the perspectives of those within them and is based upon the central premise that greater knowledge of all cultures is a worthy endeavour. As Spradley (2016) elucidates, there are five *tenets of value* that underpin this premise. The first is that understanding additional cultures can help us to explain the diversities between them, thereby aiding us to better understand humanity as a whole. The second value is that, as an inevitable effect of growing up within a particular culture, we are unwittingly trapped within its boundaries and our perceptual model of reality, our beliefs, and our values are all shaped by our culture. This is known as 'culture-bound theory', and the rationale is that, through exploration and analysis of additional cultures, it is possible to transcend our cultural bounds to understand the world from a broader and more comprehensive perspective. For us, this suggests that such research approaches could potentially expose us to new ways of thinking about, and interacting with, sound. It could enable us to expand our understanding beyond that which we have formed within the boundaries of our own culture.

The third tenet of value reinforces a point we have already discussed with regard to inductive methods. It states that ethnography is a powerful approach to grounded theory, and the construction (as opposed to the testing) of new theories and hypotheses (see Glaser et al. 1968). This is because of both the richness of the data being collected and the openness with which it is examined when using ethnographic methods. The fourth value presented by Spradley (2016) is the application of ethnography to understanding complex societies. As we move ever forward towards an interconnected and globalised society, very few cultures remain that possess only a single, clearly-defined set of cultural codes and values. Ethnography no longer serves solely as a means of better understanding other cultures, but also as a means of better understanding of our own. The implication here is that an ethnographic approach could reveal important gaps in our understanding of sound, opening new avenues for future inquiry. It may also help us begin to piece together a better understanding of 'global sound'.

The fifth and final tenet of value is to add richness to our understanding of human behaviour, specifically aspects of meaning and intent, which are both particularly challenging to ascertain without direct communication with the actor. Unlike many other branches of scientific inquiry, ethnography deals in subjects, not objects. By possession of consciousness, agency, and the ability to describe themselves and their environment, we are distinguishable from animals and other physical objects. Again, the relevance to sound is striking. Interaction with sound, the nature of the human relationship with sound, very much feels consistent with these notions of behaviour, meaning, and intent.

So, when is an ethnographic approach the most appropriate choice? As explained in Hammersley and Atkinson's primer, *Ethnography: Principles in practice* (2007, revised in 2019), the starting point for ethnographic study will typically begin with observation, description, and analysis. This will be followed by an attempt to explain what has been observed that loops back to influence how subsequent observation is to be conducted, effectively creating a continuing cycle. Ethnographic studies may alternatively originate as responses to possible limitations in a previous study, where at least one of the four links is believed to be incorrect or lacking. They may also arise in response to a surprise fact that seemingly defies immediate explanation. Alternatively, they could be stimulated by the researcher's own previous experience in which a particular phenomenon is *felt*, and potential explanations may or may not have been informally generated. One further call to arms in ethnography is the occurrence of a significant, unusual event. Such events can place people and cultures in unique positions, providing exciting opportunities to learn more about our humanness as it emerges in reaction to the event. When the Covid-19 virus became a pandemic in 2020, swathes of humanity hunkered down, severing physical interaction in favour of digital so that social distancing could be put into effect. This presented a huge stimulus for ethnographic research in general, but with a particular emphasis on interactions through digital mediators like Zoom, Facebook, and WhatsApp. In the same moment, the pandemic had a devastating impact on our capacity to utilise the majority of our methodological repertoire. This had the dual effect of drawing our attention to digital interaction as both a subject of study and a *means* of study.

Some key trends in ethnography

Earlier in this chapter, we discussed Brewer's key text, *Ethnography* (2000), and the four-link chain of social research: philosophy, procedural rules, method selection, and method execution. We also observed that, particularly within ethnography, the fundamental philosophical positions and procedural rules are now being scrutinised – not just the derived techniques, tools, and measures. This was in fact just the first of four important trends in ethnographic research observed by Brewer. The second trend is further querying of prior assumptions, in this instance challenging the distinctions between various stages of social research, and the sequence in which they are expected to be carried out. The contemporary position is that traditional research steps such as planning, data collection, analysis, writing, and dissemination may *need* to be decidedly messy. This reflects Hammersley and Atkinson's (2007) assertion that ethnographic research primarily deals not with the testing of existing hypotheses, but with steadily developing theories in a more responsive and less linear way. This resonates with the difference between Waterfall (linear, sequential) and Agile (non-linear, iterative) approaches to project management. Succinctly and poetically stated by Boehm and Turner (2003): Agile "thrives on chaos [whilst Waterfall] thrives on order" (p.58). Indeed, agile ethnography is a topic in and of itself, and has been advocated as particularly beneficial with regard to ethical concerns, as the responsiveness of the approach facilitates immediate changes in the method should any issues arise from interactions with subjects or their environment (Mara et al. 2013). The salient point here is that so much in ethnographic research is flexible and responsive, to the extent that even the questions driving the research could change beyond all recognition by the end of a project.

Brewer's (2000) third trend in social research denotes an increasing move towards research 'style' over research technique. Effectively, this move advocates bespoke methods depending on the context of the ethnographic work. It encourages researchers to identify contextual elements (that could be feasibly anything, from social media to LGBTQ+ representation, to physically dangerous fieldwork, and so on), and then to consider how the methodology and its resultant procedures and study designs should be selected and further modified to best suit these contextual elements. Finally, in a research landscape that is appearing more intricately tangled than last year's Christmas lights, Brewer's fourth trend offers some respite, noting that there is no perfect method, or even gold standard' for ethnographic research, with qualitative aspects in particular being defiant to aggressive external criticism, due to flaws and failures being especially difficult to detect and quantify. As Brewer puts it, ethnographic researchers are themselves, best-placed to critique their own approach: "they have met the enemy, and it is within themselves, for they have become their own worst enemy" (2000: p.6). This gives us, within reason, a bit of breathing space to explore a methodology and subsequent study design that makes the most sense to us and our goals. Whilst we should be considerate of the multiple levels of method, procedure, tools, and so on, we can also afford ourselves some kindness when making our selections and structures. There will be flaws, but that's okay. What is important, however, is that the methods chosen, their structure, and their execution should be clearly documented if they are to be genuinely helpful. This is especially crucial if the contextual factors are not already widely explored, as may very well be the case in our study on sound.

Autoethnography

Whilst reading through various texts on ethnography and associated research methodologies, one particular approach that caught my attention was autoethnography. Much like an autobiography, the prefix *auto* effectively denotes 'do it yourself'. A key text on the subject, *Autoethnography* (Adams et al. 2015) describes the approach as a rich self-reflection upon personal experience, meant to question the relationships between self and society from the perspective of the researcher. In their definition, autoethnography functions as a capture of a person's developing understanding of themselves, their everyday lives, and their sense of meaning within the world, with the underlying ambition that such understanding could improve the lives of others. For Adams and colleagues, the functions of autoethnography are, firstly, to retain a level of personal depth to research that is often sacrificed in pursuit of generalisability and, secondly, to leverage the researcher's ('insider') subject knowledge to provide an ethnographic account that may not be accessible by any other means. They also assert a third function, that the outputs of autoethnography should be accessible to audiences outside of an academic setting, achieved by striking a balance between intellectual and emotional writing; between methodological rigour and creative expression. To conduct an autoethnographic study, Adams and colleagues suggest the following:

1   Foreground personal experience in research and writing
2   Illustrate sense-making processes
3   Use and show reflexivity
4   Illustrate insider knowledge of a cultural phenomenon/experience
5   Describe and critique cultural norms, experiences, and practices
6   Seek responses from audiences.

(Adams et al. 2015: p.26)

False modesty aside, I feel it may be a little self-aggrandising to suggest that this book has the potential to improve the lives of others, though ultimately that is the hope, however minor the effect may be. In a subsequent article on the subject, Adams and colleagues (2017) each provide personal examples of their own autoethnographic work that clearly positions the purpose of such an approach in contexts concerning highly sensitive issues relevant to potentially vulnerable groups within society. As such, I don't feel it even appropriate to suggest that conducting a personal account around the matters of my everyday interactions with sound can fully qualify as autoethnographic. However, the balancing of emotive and analytic content certainly resonates with any ambition I may have to make this book accessible and engaging, making the borrowing of certain autoethnographic study features an interesting proposition.

The above detail is admittedly rather abstract, and I have not yet provided a sense of precisely how an autoethnography could be conducted that would be compatible with our investigation into the human relationship with sound. We shall return to this shortly.

Is ethnography the right choice for studying sound?

In review of the various discussions above, there is most certainly a strong rationale for utilising ethnographic research strategies. There is little room to argue against this rationale because our research ambitions are most definitely relevant to raising awareness of diversity; overcoming cultural bounds; generating new theories and hypotheses; understanding cultural complexity; and learning more about human intent, behaviour, and our sense of meaning about sound. These are very much our aspirations, but this is most certainly not the first project to have chosen ethnographic methods for studying sound. We now turn our attention to sound studies and the strong historical precedent we have for selecting ethnography.

Making sense of sound studies

As a subject of study, researchers have been considerate of sound to greatly varying degrees. Often, thinking upon sound has found itself embedded within other topics and disciplines. Historically, it has not been an exclusive subject in itself. In the first few years of the 21st century, researchers observed that the term 'sound studies' may have entered our academic vocabulary but was yet to be properly defined and formalised as a research discipline (Bull & Back 2003). We shall take a closer look at sound studies shortly. First, a step back in time with a brief overview of a close relative, ethnomusicology.

Ethnomusicology, for a bit of history

Broadly defined as the study of music in its social and cultural contexts, ethnomusicology was more widely acknowledged as an academic discipline from the early 1990s (Schuursma 1992) but dates back several decades prior to the mid-20th century. This gives it a significantly longer documented history compared to sound studies. Articles published around the time of ethnomusicology's formal conception, described it as the study of music-as-phenomenon, encompassing physical, psychological, aesthetic, and cultural domains (Hood 1957). The Society for Ethnomusicology[1] outlines three core features of the discipline: a global perspective, music as social practice, and interdisciplinarity with an emphasis on ethnographic and historical methodologies. Rice observes that ethnomusicology is, for some, a theory-less subject of study that represents a "domain of interest shared by a community of scholars" (1996: p.102), whilst for others, it is formed of a substantial array of theories but little theoretical consensus and certainly no uniting paradigms. For Post (2004), ethnomusicology embodies many characteristics discussed earlier in this chapter. Post's work features terms such as comparative, human expression, everyday life contexts, and cultural events. On the surface then, it does appear as though ethnomusicology reflects key values of ethnography, with an emphasis on music as the object of study and human culture as the subject.

In a highly candid essay, Wong (2014) helps us to understand the nature of ethnomusicology through the steep hills it has been made to climb on its journey to genuine acceptance within Western musical discourse. Wong stresses the contrast between a logocentric Western understanding of music and a relativist ethnomusicological understanding. The Western tradition holds great value in the words and symbols with which it represents and expresses music, positing them as accurate and consistently objective. Ethnomusicology rejects this objectivity almost entirely, embracing the position that the nature of music is relative to the perspective of the listener and the context within which the listening is occurring. Reading Wong's essay brought to mind a memory from my early career years as a secondary school music teacher and a conversation I had with a colleague that has somehow become etched in the back of my mind. The conversation was regarding appropriate performance technique when playing the piano score for Steven Sondheim's *Sweeny Todd: The Demon Barber of Fleet Street*. Coming from more of a popular/jazz background, I was used to playing by ear and performing through a blending of more abstract direction from the composer and improvisation from myself. I innocently (and rather naively) asked if it mattered whether or not I played every note precisely as written. Then I wished I hadn't. The response I received was an 'education'. To put it mildly, I was encouraged to "*fully appreciate the value of the score as written*".

Hopefully, a reassurance to Wong, Harrison (2012) asserts that the stock in ethnomusicology appears to have been on a steep rise over the past decade, particularly as an applied study. Harrison's review on the knowledge-generation principles of applied ethnomusicology emphasises a focus on the role of music within a social organisation. Specifically, it explores ways in which people interact with the environment and with each other through music. It also strives to uncover their motivations for doing so and to better understand the nature of their experience. This indicates that ethnomusicology is progressing, giving us further confidence in the use of ethnography as applied to non-musical sound.

Ethnomusicology also draws sharp parallels between music and ethnographic research methodology that I would summarise in the term 'improvisation'. Historically, both music and scientific inquiry held on passionately, some may say dogmatically, to their rules and procedures, believing the foundations of these to be absolute. Some may say sacred. Then, encroaching on this hallowed ground, stepped the improvisors, intent on questioning everything, starting from scratch with even the most foundational aspects of methodology. Inspired on a whim to type "ethnology is the jazz of science" into Google, I was returned numerous published texts dating back several years, depicting the ethnographer as equal parts jazz soloist and research scientist (Humphreys et al. 2003; Solis 2014). This wonderfully metaphorical blurring really does reinforce what I would argue to be the central theme that has emerged from this chapter. If we characterise sound, music, and our understanding of culture, society, and relationships within human experience all as a singular composite entity, that entity is far too rich and downright labyrinthine to be examined with fixed sets of preconceived approaches and expectations. We are compelled to explore the subject, our processes, and ourselves as researchers all in parallel.

Does sound studies actually study sound?

Eloquently described by Sterne (2012), sound studies "redescribes what sound does in the human world, and what humans do in the sonic world" (p.2). Sound studies considers the nature of sound both as an object and as a subject. In terms of the former, sound studies research distinguishes between four key sound object types: sound, music, noise, and silence (Pinch & Bijsterveld 2004). However, it is sound as a subject of study that most research within the discipline is interested in. Sound studies takes a broad, inclusive, and open-minded approach. It is not the study of sound in isolation, but rather the nature of sound within culture and society. Kelman (2010) eloquently summarises the motivation underpinning the field as an interest in "[...] understanding how sound circulates and how it contributes to the ways in which we understand the world around us. In other words, [...] the relationship between sound and the social production of meaning" (p.215).

When we consider both Kelman's and Sterne's descriptions of sound studies with reference to our earlier discussions in this chapter, there is certainly a distinct echo of ethnography here. Surprisingly, the explicit connection between ethnography and sound studies is decidedly new, despite a central tenet of the former being the value of listening. In the last few years however, various researchers have started joining the dots. For anyone interested in digging deeper into the discipline of sound studies, Bull's *The Routledge Companion to Sound Studies* (2018) serves as a comprehensive introduction to the subject. As an edited volume, the structure of this book gives insight into the key components of the discipline, many of which are distinctly ethnographic in their focus. Individual chapters address the nature of sound as it relates to matters of gender, politics, racial difference, geography and place, history and archaeology, social interaction, aesthetics, and culture. Here, the table of contents, whilst admittedly a superficial means of observation, already hints at sound studies being dramatically relativist and multidisciplinary. With a resolute kick-start to the volume, Grimshaw-Aagaard (2018) questions the fundamentals of what is (and moreover, what *should be*) understood as sound studies. In this chapter, one particular position presents a methodological challenge that is becoming something of a recurring theme – that, because there remain vital questions upon precisely what sound is, the methodologies, procedural rules, and resultant study design approaches can all be subject to questioning and should not be taken for granted. This ultimately reinforces our earlier concern that any decisions made in terms of method for this book can only be justified to a limited extent. As such, this book is far more a grounded theory project; synthesising useful questions rather than definite answers is a far more appropriate ambition.

Grimshaw-Aagaard's depictions of the discipline reflect our earlier observation that sound studies are primarily concerned with the examination of auditory culture. This gives sound studies undeniably ethnographic overtones. This chapter also asserts that science and technology have a history of connection to sound studies, both as broad forms of study and as methodological sources. This emphasises the truly multidisciplinary nature of sound studies – sitting at the crossroads of science and art, technology and design, physiology and psychology – all embodied within a cultural wrapping of politics, history, sociology, economics, anthropology, and the list goes on. In terms of technology, sound studies is primarily interested in human interactions, specifically, the means by which we capture, store, measure, manipulate, transfer, and conceptualise sound and music. What sound studies largely appears to avoid, paradoxically, is the study of what some may call sound itself, but what we shall refer to as 'acoustics'. Grimshaw-Aagaard remarks that this avoidance is likely an issue intertwined with the dominance of the physical interpretation of sound (discussed in detail in the previous chapter), and objection to the ensuing assumption that "sound is the province of acousticians" (p.18).

In service of full transparency, I should note that Mark Grimshaw-Aagaard was my PhD supervisor and we have collaborated on writing projects on and off for several years. So, of course there is a little personal bias here, as the notion of sound as an undetermined, emergent phenomenon was part of my research upbringing. That said, there is a clear consistency between this position on the questioning of fundamentals of sound studies and the trends observed across ethnography and ethnomusicology. The jazz musicians have entered the building.

Approaches to sound studies

During the course of writing this chapter, the preeminent writer upon sound as acoustic ecology, Raymond Murray Schafer, sadly passed away. In his wonderfully accessible text, *The Book of Noise* (1970), Schafer instructs us to "[l] isten carefully with seismographic delicacy to the sounds of the environment around you. Close your eyes and listen with musicians' ears (just for 5 minutes)" (p.3). Schafer then asks several questions of us as listeners. These questions should resonate significantly, both with the overarching aims of this book and with the ethnographic and sound studies principles that we have discussed so far:

- — What was the loudest sound you heard?
- — What was the softest?
- — What was the highest?
- — What was the lowest?
- — What was the most beautiful sound you heard?
- — What was the most unbeautiful?
- — If you could, what would you change about the acoustical environment?

(Taken from *The Book of Noise*, R. M. Schafer 1970)

For Rice (2018), an 'ethnography of sound' functions as a means of revealing local sonic forms and uncovering the relationship between those sonic forms and the society within which they occur. It is meant to draw clearer connections between the contextualising factors of the local *habitus* and how we make, receive, and interpret sound. As Rice explains, with both sound studies and ethnography being relatively young disciplines that characteristically defy fixed definitions, the methodology for sound ethnography suffers the same difficulty twice over. As such, there is very little information out there to direct us in formulating our own method. This provides us with a nice amount of freedom but at the cost of limiting our confidence in whatever approach we decide to take. Rice's essay provides the most appropriate form of guidance under such circumstances by detailing several examples of sound ethnography. Through these examples, several points emerge that resonate with certain specifics of the ethnographic method. In describing Feld's (2015) fieldwork with the Kaluli of the Bosavi rainforest, Papua New Guinea, the notion of ethnographer as learner and subject as expert is emphasised. Feld's work also reveals interesting interactions with audio recording technology as a means of acquiring data. Members of the Kaluli would direct Feld in his microphone placement and then tweak the mix of the captured soundscape to better reflect their perception of the dimensionality of the sound. There is a wonderful sense of learning about sound from people and at the same moment, learning about people from sound.

Recounting his own research into the soundscapes of hospitals, Rice presses upon the importance of what he calls 'situated listening'. The purpose of situated listening is to personally experience sound in a way that is as close in situational context (location, time of day, proximity to key events/sources, duration of exposure, physicality, personal circumstance) as that experienced by the ethnographic subject of the study. There are, of course, unavoidable limitations to situated listening as Rice points out, referencing the assertions of Chandola (2012), whose fieldwork in the deprived settlements of Delhi led her to realise that despite being present in the space for extended periods of time, she could not discount her own middle-class status as a barrier to being fully situated[2].

Ultimately, and this goes primarily to the individual at my university in charge of granting travel budgets, it appears as though a round-the-world trip is very much in order. Of course, prior to (or more likely, in lieu of) that, it is worth noting that a great number of soundscapes from around the world have already been captured and catalogued, in one set, by the British Library (BL)[3]. The BL collection of soundscapes are alphabetically organised by location, country, then region (the specificity of which depends on the size of the collection for that country). As you might expect within a BL collection, the captures from England are notably greater in number than those from other countries. I would also like to point out that the regions featured within England do appear to heavily favour the South. Where are all the Yorkshire-based sound-recordists hiding? Putting the North-South soundscape divide aside, the BL soundscape collection of England appears to reflect the notion of 'place' more broadly. Though the vast majority are wildlife and nature soundscapes, the captures tend to emphasise a single sonic environment that epitomises that particular locale. For example, Cornwall features several captures taken on beaches. Hampshire's collection by contrast favours woodland, forests, and harbours, whilst Norfolk presents wetlands and birdsong. London's recordings stand out, both for being significantly more urban and also by being largely taken from atop a moving Routemaster open-topped tourist bus, presenting a dynamic capture of the city soundscape from a point in motion. You could of course retort here by suggesting that knowing the type of soundscapes that were deemed to be exemplifiers of a particular place absolutely provides us with some ethnographic insight by indicating a value judgement on what *should* be captured from a certain locale. I would completely agree, however the remit of this book aims to be more than a little wider but also, perhaps more importantly, more 'everyday', and I'd wager few of us spend the majority of our everyday lives atop a tourist bus, strolling through wetlands or sitting on a beach. Sadly.

We explore soundscape recording in greater detail in Chapter 5, in which the focus is on sound and place, but it feels appropriate to give a brief overview here. Soundscape recordings can be static captures, using one or more microphones in a fixed position, or dynamic captures, such as the Routemaster London bus recordings. The latter is a variation on the theme of 'soundwalks', in which the recordist creates a dynamic capture as they travel along a pre-determined route around a locale (see Liu et al. 2014). Audiovisual equivalents of this are notably popular in online culture. Examples of this include *Nomadic Ambience* and *Keezi Walks* – both channels on YouTube. Channels like these can boast hundreds of thousands of subscribers and can accumulate millions of views in some instances. Of course, these videos exist primarily for entertainment purposes, but there is arguably a good degree of insight that could be gained from a structured analysis of the soundscapes in these walks.

Doing so is completely out of scope for this book; however, after viewing more than a few of these in full, I would recommend to you the uncanny near-silence of a snow-drenched Central Park in New York City, USA[4], or the cacophonic motorcycle chorus of Lahore, Pakistan[5].

Whilst sound-capture studies such as those by Feld (2015) and Chandola (2012) realise ethnographic depictions of *other* people and places, and the various online soundwalks broadly exist to serve a similar function, there is the opportunity here to repurpose such techniques in service of an autoethnography. As I alluded to a good few pages prior, a practical means of conducting an autoethnographic study is something that I was confident would have value, both in prepping for the interviews and in complimenting the contributor responses. Considering the above discussion, situated listening and soundwalks appear to be a good solution. Referring back to Adams and colleagues' (2015) recommendations for an effective autoethnography, reflective writing on captures taken from personal soundwalks around my home of Portsmouth would most certainly illustrate sense-making processes as I described my interpretation of the many sounds around me. I would be able to frame the writing with reference to, where relevant, the various theories and perspectives detailed in Chapter 1, thereby fulfilling the requirement for illustrating 'insider knowledge'. Lastly, such recordings would arguably be ideal stimuli for prompting descriptions and critiques of cultural norms, experiences, and practices.

The wider value of soundwalks features notably within academic literature. Polli (2012) for example, observes that "the act of listening through public soundwalks and other formal and informal exercises builds environmental and social awareness and promotes changes in social and cultural practices" (p.257). The idea of engaging in a soundwalk is indeed an intriguing proposition. What hidden elements within my daily life might I notice when attuning my attention exclusively to the soundscape? What could my perceptions, impressions, and affectations towards these sounds tell me about myself?

To summarise, sound studies research is arguably progressing at a rapid pace, though it remains a fledgling discipline. As such, its methods are continuing to develop, and this is something of a long game. One of the broad agreements between researchers is that sound studies is multidisciplinary, with many of the fields that feed into it being subject to ongoing methodological development. For a newcomer, it's like trying to put together a 10,000-piece jigsaw puzzle where, not only does the overall image keep shifting in appearance, but the shape and connectivity of each individual piece also keep changing. The solution appears to be to avoid putting all your methodological eggs in one basket and instead combine multiple acquisition and analytical tools together, allowing the strengths of one to compensate for the limitation of the other.

So, you want to do some interviews?

Interviews are pretty popular. In the dissertations and postgraduate theses of my student supervisees, they come in second only to the questionnaire.

Whilst I absolutely acknowledge the value of that particular research method, I would be prepared to make any number of highly dubious deals with the pan-dimensional lords of academia, to ensure that I *never* see another student thesis where the sole source of data is a questionnaire ever again. Interviews are, or at least can be in some regards, a close relative of the questionnaire, particularly if we agree with Maccoby and Maccoby's (1954) historical view of an interview's function as "attempts to elicit information or expressions of opinion or belief from another person or persons" (p.449). A review by Young and colleagues (2018) describes how the use of interviews stretches back many hundreds of years, first finding formal scientific usage towards the end of the 19th century. Their writing emphasises the role of mutual learning, where a successful interview is one where the researcher may find answers to their initial questions, but also uncovers information on perspectives, themes, or topics highlighted by the interviewee, that the researcher never thought to question.

Returning briefly to Saunders and colleagues' (2009) research onion, it is worth noting that there is clearly no exclusive route from epistemological philosophy to the use of interview as a procedure. The interview appears mostly free to feature across numerous types of study design. The rationale for selecting an interview is highly likely to be qualitative, inductive, and based upon interpretivist and pragmatist philosophy. That said, it is certainly possible for an interview (or parts of one) to be more empirical, to deal with quantitative data, and to be made sense of through a more deductive process. This means an interview could be part of an experiment or be the basis of a survey, or feature across almost any of the strategies identified in our beloved onion. They could be the exclusive method, or they could be combined with others. They could be cross-sectional or longitudinal, delivered just once or at multiple points over a given time period. What this ultimately means is that there is no top-level diktat on whether an interview is appropriate for a particular study. The devil is in the detail.

When and how to use an interview

So, how do we know using interviews is a good choice for our purposes? For Kvale (1996), interviews should be deployed when the aim is to uncover both factual and interpretive information. If we need to know about something, but also what that something means to the participant, then we have a reason to select an interview. Valenzuela and Shrivastava (2002) observe value in interviews when the opportunity to further probe participant responses and collect large volumes of rich data without putting significant demands upon the participant is required. In contexts such as ours, there is a priority need to understand both factual and interpretative information (and the relationships between the two), plus very little in the way of pre-established hypotheses, and no clear rationale for prescribed questioning. These points strongly align with both Kvale, and Valenzuela and Shrivastava's descriptions of circumstance in which an interview has a distinct value. So, if we're in agreement that the humble interview is looking like a decent option, how do we design and deliver a *good* interview? Young and colleagues (2018) provide us with nine basic stages of interview design and delivery:

1 Identification of research question
2 Selection of interview type
3 Generation of initial interview questions
4 Identification and recruitment of interviewees
5 Ethical review
6 Pilot interviews to refine the design and delivery
7 Undertake interviews
8 Analyse results
9 Write-up: conclude on results and formulate or update theory.

(Young et al. 2018)

Whilst we haven't formulated a final list of questions at this stage, our review in Chapter 1 did help us to set various initial questions and, as both Hammersley and Atkinson (2007) and Brewer (2000) kindly reassured us earlier, ethnographic research methods are fully supportive of flexible study designs that are responsive to methodological insights gained during the actual process of data collection. We're free to change our questions as we go, should we feel the need. This brings us neatly on to stage two: selection of interview type. Bauman and Adair (1992) identify and compare five types of qualitative interview, relevant to social research: (1) in-depth, unstructured, non-standardised; (2) in-depth, structured; (3) focussed; (4) psychological–clinical; and (5) ethnographic. The various overlaps and differences between these groups are easier to discern in a more at-a-glance form, so let us treat ourselves to a nice table.

Using Table 2.1, it isn't too difficult to start pruning away alternative interview options. Testing hypotheses rather than generating them takes Psychological-clinical out of the running. Next, we cut Focussed for requiring pre-defined questions and de-emphasising wider context. That leaves us with Ethnographic, In-depth/structured, and In-depth/unstructured. How the nuances between these options are perceived could be subject to debate but, in considering the options myself, the key factor affecting my choice was on the matter of the interviewer directing the conversation. Unstructured placed the greatest emphasis on interviewee-led discussion, with structured being interviewer-led and ethnographic somewhere in between the two. This really did feel like a compromise either way. If I directed the conversation, I could have more power to produce transcripts that could be directly compared between contributors but would potentially miss out on unexpected topics or perspectives that I hadn't even considered asking about. Alternatively, I could allow the interviewee total freedom and remain open to surprising findings but compromising with the risk that everyone would explore

Table 2.1 Broad comparison of qualitative interview types (based on Bauman & Adair 1992)

Interview type	Open questions	Pre-defined questions	Imposed topics	Set order	Key features
In-depth/ unstructured/ non-standardised	✓	✗	✗	✗	– Conversational – Provocative questions – Requests for explanation/ elaboration
In-depth/ structured	✓	✗	✓	✗	– Meaning coded by researcher – Interview is guided but questions are flexible in wording and delivery
Focussed	✓	✓	✓	✓	– Employs stimuli to trigger responses – Codes four response data types: range, specificity, depth, and personal context
Psychological-clinical	Mixed	✓	✓	✓	– Questions formed from an initial hypothesis – Questioning flexible to test the validity of the hypothesis or generate alternative
Ethnographic	✓	✗	✗	✗	– Topics can be steered to but not imposed – Interviewee as expert, not researcher – Focus on everyday experience/point of view – Language as data

totally different topics, thereby minimising the opportunity to draw valid comparisons. Not willing to accept the full consequences of either, I chose Ethnographic under the hope that this would be a balanced compromise but would also enable me to manoeuvre more towards structured or unstructured throughout the course of the interviews – should experience of the interviews themselves reveal a need to do so.

In their review of interview technique, Valenzuela and Shrivastava (2002) offer a helpful primer for budding interviewers. Within their guidance, they emphasise the qualities of an effective interviewer and also provide a taxonomy of information types that is particularly helpful when structuring an interview and selecting topics and individual questions within. On being an effective interviewer, they advise: being knowledgeable on the subject of discussion; clearly outlining the interview procedure and asking clear, intelligible questions; being tolerant and sensitive to potentially provocative and discomforting opinions; being able to steer the interview and avoid digression; and effectively interpreting responses to support responsive follow-up questioning. On the matter of information types, these are classed as that which is relevant to: action (behaviours); cognition (thoughts, opinions, and values); emotion (feelings); knowledge (factual information); recollection of sensory experience; and background/demographic data. Of course, it may not be necessary to collect information across every one of these types, but it makes sense to consider what types we are collecting, and the relative volume of information was collecting per-type, against our foundational research focus of the human relationship with sound – asking if the features of the former are the best way of answering the latter.

The ethnographic interview

We have already covered issues surrounding interviews, their purpose, value, and how and when to best deploy them. However, based on the details on ethnography explored above, it now makes sense to focus our attention on our final selection, the ethnographic interview, to see if it really is the best instrument for our purposes. As an ethnographic research instrument, interviews fall under oral accounts as an overarching set of tools. Oral accounts may be solicited or unsolicited, the former providing far greater researcher control at the cost of attaching substantial researcher bias and likely, if inadvertently, closing the door on potentially pivotal information, whilst the latter presents the opposite conundrum, affording more naturally occurring and potentially richer information at the cost of likely constraints to the relevance of what is said against the ambitions of the research. Interestingly, when collecting oral accounts, it is possible for a researcher to receive unsolicited information even in a structured interview context, in scenarios where the participant feels the need to go beyond the scope of a question due to their own agenda.

For Hammersley and Atkinson (2007), an ethnographic interview can be scrutinised to serve two functions. The first is quite simply to inform us of the phenomena they describe. The second is to elucidate upon the perspectives, discursive strategies, and psychosocial dynamics that they imply. Whilst perspectives can denote a very wide range of things, discursive strategies describe the use of various linguistic techniques that broadly support the speaker in presenting themselves more positively, or possibly presenting others more negatively (Ramanathan et al. 2020). Examples of discursive strategy include predication (describing other individuals or groups positively or negatively in a social context, often using a metaphor), argumentation (attempts to justify an act that had a negative outcome, or exaggerate a positive outcome, by framing it in a particular context), and referential nomination (categorising people into 'in' and 'out' groups to create a sense of 'us and them'). Psychosocial dynamics is a term that often pops up in developmental psychology articles, particularly those addressing adolescence. It describes ways in which the cultural environment and social conditions within can affect an individual's physiology, their psychological state, and their resultant behavioural, emotional, and perceptual responses that can, in turn, feedback into the environment through further social interaction (Eckersley 2016). Though of course, not everything a participant says can be taken at face value, a well-designed and executed interview can realistically have ambitions to generate information across these psychosocial components. We may ask a participant about their behaviours but also the situational factors that stimulated their actions, plus the emotional and cognitive processes that formed before, during, and after.

Being written specifically to inform on ethnographic practice, Hammersley and Atkinson's (2007) work is difficult to avoid citing when discussing the important things to consider when preparing an interview. The following is a concise paraphrasing of some of their key recommendations on the subject:

-   Interview questions may need to be avoided if there is the risk of them being perceived to be invasive, aggressive, or threatening.
-   Naturalism can be a potentially vital issue in avoiding interview data as 'co-construction' between interviewer and interviewee. Non-directive interviewing and open questions can help mitigate this.
-   Establishing rapport is broadly desirable, particularly in circumstances where the interviewer is looking to obtain personal and possibly sensitive information or is asking the interviewee about matters that they may not be immediately comfortable discussing. One potential strategy is interviewer self-disclosure.
-   An ethnographic interview is typically light on both interviewer-control, constraints, and formality. It is inaccurate to describe them as unstructured because they absolutely are, though not exclusively by the interviewer, more so collaboratively structured by both interviewer and interviewee. Whilst there may be a degree of direction from the interviewer, there is typically little predetermined structure. The tone is conversational and there is an emphasis on relaxed open, dialogue with reflexive, often branching questions from the interviewer, but also the space for questions from the interviewee.
-   Interview preparation commonly extends to a list of issues to be covered, but no sequence of questions.
-   Questions may be non-directive or directive (requiring a specific response or not, respectively) and this can be decided in real-time as the conversation progresses.
-   Interviewees in their own 'territory', in a meeting organised in a way that suits them is broadly acknowledged to be beneficial for rapport, trust and comfort.

(Based on Hammersley & Atkinson 2007)

Analysing interview results

Checking off the many items on our expedition prep to-do list we have just two more things to consider. The first of these is analysis. Hammersley and Atkinson (2007), once more, provide us with some reassurance when making decisions on how we approach analysis, stating that "there is no formula or recipe for the analysis of ethnographic data [and] certainly no procedures that will guarantee success" (p.158). Indeed, Hammersley and Atkinson make it abundantly clear that anyone prescribing fixed recipes to ethnographic method should be roundly ignored. This does not mean, however, that all of more high-level or broader aspects of methodology that affect analysis are also immune to guidance. As an inductive process, ethnographic research seeks to build a new theory, not to test the existing theory. Therefore, analytical techniques that support 'grounded theory' come highly recommended. Grounded theory largely supports the requirements of ethnographic studies to acquire data naturalistically (i.e. almost invisibly, without significantly impacting upon the environment) and in a reflexive way. The trade-off here is that natural and reflexive acquisition is very difficult to control, so acquiring a data set with various elements that can be validly compared (such as comparing one interviewee's responses to the next) can be difficult, if not impossible in some cases. This reflects my earlier concern in the significant trade-off between richness and control when comparing structured with unstructured interviews. In terms of analysis though, this often means that researchers process their data descriptively, extracting interesting observations and broad, overarching trends.

To some extent, many of the points above give the impression that we can effectively make things up as we merrily go along. As Burns (2021) explains, the improvisatory nature of ethnography dictates rather messy analysis, but continues, noting that the process can still be readily broken down into three steps: coding, thematic analysis, and theorisation. Coding kicks things off as a mechanism for making some initial sense of our data by assigning words or phrases as labels to individual pieces of data. As we are working within an inductive methodological structure, inductive coding means that we do not start the process with any pre-defined labels. These labels are generated and applied in parallel, as we go through our data, point by point. Guidance on coding is quick to point out the labour intensiveness of this process, with inductive coding requiring several iterative revisions, in which you apply as many labels as you can on a first review of the data, then evaluate the overall cohesion and usefulness of your coding framework to decide what labels need adding, removing, or editing (Burns 2021; Medelyan 2021). This process can require several iterative cycles. If you have a large dataset, I pity you. Unless you have a postdoc to do this for you. In which case, well, it's tantamount to abuse but good for you.

Once we have coding satisfactorily completed, we then move on to thematic content analysis (see Brewer (2000) for a comprehensive review). As the name suggests, thematic content analysis involves structuring our newly coded data into classes that can themselves be codified to become themes of our data. Depending upon the complexity of the data, multiple layers can be formed by further grouping underlying themes into overarching or compound themes. This is what's commonly described as a bottom-up process, in that we start with small units of data that are used to build larger thematic structures, but also in that we start with low conceptual level (concrete) data and use that to construct high conceptual level (abstract) information. This higher-level information is then considered against the relevant academic literature as we move from thematic analysis to theorisation. This is our complete process for translating observation into theory.

Once again, re-treading the now well-worn path of ethnography's distaste for heavily prescribed approaches, thematic content analysis, and the theorisation stage that follows, are both open to interpretation and improvisation. Researchers are encouraged to produce a bespoke recipe rather than follow a generalised template, if such a thing even exists (Madden 2017). However, broad guidance is still available and worth bearing in mind for the shell of our approach, with Burns' (2021) online review providing several highly useful 'ingredients' to consider:

- *Frequency analysis:* How many times certain codes or themes appear within the dataset. Observing high frequency between participants could indicate generalisable trends/consistencies whilst high frequency within a single participant's data may suggest deeper relevance and personal impact.
- *Hierarchy of significance:* Considering relationships between multiple themes may imply correlation or potentially causal links. The presence and strength of these relationships (perceived significance) can be tagged, and these features can themselves be compared and their relative strength can be used to build a hierarchy. Significance can be attributed based on...
  - ...*absence:* Looking for gaps or outliers in the data; circumstances in which you were expecting a particular response, or data that fed into an anticipated theme, but such points simply did not occur. This could be absence across the entire dataset or between individual participants.
  - ...*expression:* identifying certain codified data as more meaningful when considered against the emotional intensity with which it was expressed.
  - ...*critical incidents:* attributing greater importance to points that are presented as the core of a participant's story or anecdote.
  - ...*relevance to the research question(s):* tagging data points, trends, and overarching themes with relative relevance scores.
- *Assertion development:* Considering all aspects of the data, and the narrative of its creation, against relevant academic literature.

Highlighting anticipated and unexpected findings, patterns at multiple conceptual levels and overall conclusions that then form the
final theory.

(Based on Burns 2021)

How can technology help?

Before we close this chapter, it is worth mentioning one final element – an
element that resides more or less in the heart of the research onion as part
of the techniques and procedures layer within which we also found analysis.
The joys of exploiting postdoctoral research assistants aside (just to be clear,
no research assistants were harmed in the making of this book), thematic
content analysis is unquestionably a long and drawn-out process. Fortunately,
computer software does have a decent propensity for addressing matters of
efficiency, making mere humans redundant in the process. Its response to the
challenge of thematic content analysis is qualitative analytical software. These
programs utilise natural language processing and machine learning technologies to automate the analytical process. As Medelyan (2021) observes, there
are pros and cons to using such software. The efficiency advantages are the
standout benefit, but using an automated system also provides scalability. It
has further been argued that using software also helps by avoiding researcher
bias. Medelyan counters these benefits with limitations in accuracy, noting
that a human researcher is able to draw out finer details and more reliably
identify patterns than a machine is (currently) able to do. For our purposes,
this comes down to a matter of scale. If the dataset is unworkable for one poor
soul to process manually, then off to the machines it will merrily go. Where
the line is between workable and unworkable scale is up to me, and I could
use more sleep as it is.

Despite a wide range of technologies available, contemporary ethnographic research (particularly cross-cultural studies) broadly continues to favour relatively low-tech audio recording alongside pen and paper (Kahn et al.
2019). The reasons for doing so primarily come down to the lower risk of
equipment malfunction or researcher error when dealing with unfamiliar or
complex electronic technologies, but also concerns with data security when
capturing large volumes of data electronically, that will not be automatically
filtered for overtly sensitive information as would be possible with pen and
paper (Wutich & Brewis 2019). Participatory techniques that engage subjects
in various activities as part of an ethnographic study (such as brainstorming
or affinity mapping) also heavily favour low-tech pen and paper, featuring
good old chunky markers and post-it notes (Tickle 2017). Of course, in the
face of the Covid-19 pandemic, qualitative researchers quickly discovered a
need to pivot towards online technologies as the ability to conduct face-to-
face research was obliterated almost overnight. At the time of writing up the
final draft of the methodology, the official UK Government position on the

pandemic was to encourage a 'managed return to normal', meaning univer-
sities were, mostly, able to revisit in-person research methods. That said, the
near-24 months of lockdowns and restrictions contributed to a flurry of studies evaluating the potential effectiveness and limitations of online methods.
As Howlett put it well, we are "[l]ooking at the 'field' through a Zoom lens"
(2021: p.1). Howlett's review identifies several positives in online technology
for ethnographic interview. Early findings have suggested that mediating a
conversation through a pair of screens can actually reduce participant discomfort and encourage them to discuss personal issues more freely. Online
methods also provide convenient access to both audio and visual capture of
the interviewee and provide access to a much larger network/pool of potential participants. They can also facilitate much easier international collabo-
ration between researchers or enable a focus group to contain people from
around the world far more easily than if trying to achieve the equivalent in a
shared physical space.

No single technological solution is without fault, and recent research has
also raised concerns that videoconferencing methods may be skewing findings towards advantaged cohorts that have access to the required computer
hardware and online infrastructure (Kennedy et al. 2021). In a review of
numerous quantitative and qualitative social science methods, Hensen and
colleagues (2021) discuss videoconferencing and reiterate the issue of the
technology costs prohibiting access to certain demographics. They also argue
that the 'veil of technology' also reduces the researcher's potential to develop
rapport and establish trust. Their review further identifies several logistical
challenges, such as disturbance from unwanted noise and potential reliability issues with connectivity, battery life, and bandwidth. Two particularly
concerning issues are privacy and data security, with Hensen and colleagues
arguing that videoconferencing can have limitations in terms of end-to-end
encryption, but also places the responsibility for privacy entirely upon the interviewee and, much like access to the technology, favours more advantaged
individuals – so too does personal access to a quiet and private space. On the
face of it, conducting interviews online has significant logistical benefits,
allowing us to cast our net wider and capture richer data very efficiently.
Issues with privacy, data security, hardware access, and reliability need to be
accounted for but, particularly bearing in mind the current climate, the use
of this technology is quite frankly, unavoidable.

Within broader qualitative and social science research, it seems logical to
suggest that the subject or topic of an interview is unlikely to have much
impact on the technology being deployed. However, field research within
sound studies frequently uses various pieces of recording technology to capture auditory content. This is especially commonplace in soundwalks that we
discussed earlier. Based on examples of these soundwalk studies, head-worn
binaural microphones appear a highly suitable recording solution (Butler 2007;
Jeon et al. 2014). This does make a great deal of sense as these devices can
capture much of the spatial quality of the soundwalk. They are lightweight

and comfortable, making them very portable as part of a moving recording. They are also relatively discrete, enabling the researcher to capture the environment without drawing the kind of attention you'd invariably get wandering around a well-populated urban space, your arms outstretched, carrying a traditional outdoor microphone – or, if you really want people to stare, a binaural dummy head, or even (my personal favourite) an 'omni-binaural' microphone rig, which is basically a bunch of ears on a stick[6].

The method for this book

If you ask questions on methodology, particularly how to design a *good* ethnographic study, guidance can range from following various heavily prescribed and fixed doctrines to being told to 'get out there and figure it out for yourself' (Hammersley & Atkinson 2007). The 'classical or jazz' dichotomy is certainly a challenge, as aligning with one perspective will inevitably draw criticism from those who advocate the other. All things considered, however, it does seem as though there simply isn't the required evidence base to justify embracing a more fixed and micromanaged study design. At the same time, trends in relevant research domains largely appear to be moving further away from such methods, not closer to them, as more evidence is gathered. This suggests that the ethnographic community is increasingly attributing more value to the 'jazz' way of doing things.

So, to you, the intrepid reader who made it through this chapter, I commend you and hope that you feel joining me in this methodological expedition was worth your time. Based on everything that we have discussed above, the following is a list of the key methodological features that were employed to produce the rest of this book:

1 *A non-prescribed methodology*: The underlying philosophical positions, resultant rulesets and specific approaches, and study designs are all subject to questioning in the context of an investigation's specific aims. This provides academic freedom but requires the researcher to fully explain their approach and generate appropriate data to help comprehensively evaluate its effectiveness.
2 *An inductive process*: We are looking to generate theory, not to test it. Therefore, inductive approaches are preferred over deductive.
3 *An interpretive approach with qualitative data*: We are dealing with matters of meaning and personal experience. As a result, interpretive approaches that generate qualitative data are essential to capturing the required nuance and richness.
4 *Mixed methods*: As an exploratory work, focussing on a singular method would be an 'all your eggs in one basket' move. A poor choice of method would have disastrous consequences in this scenario, and it would limit the potential to produce a wide range of new insights into methodological process. Therefore, a mixed-methods approach is needed.

5 *The interview*: Direct observation in situ across multiple countries is un-
feasible, whilst questionnaires lack flexibility and intimacy and do not
support responsive questioning to capture unexpected findings or pursue
new lines of inquiry in real time. An interview supports these requirements very well, whilst also being more logistically manageable and cost
effective.

- *Ethnographic interview*: This technique best reflects our aims, featuring: topic steering (but not imposing), open questions, focus on
  everyday experience, consideration of language as valuable data, and
  viewing the interviewee as an expert of their own interpretation of
  sound.
- *Five-feature dataset*: The interview should aim to capture five features
  from interviewee accounts: sensory experience, behaviours, cognition, emotion, and knowledge.
- *Sound-relationships to capture in the data*: To cover the relationship between sonic forms and their locales comprehensively, questioning
  should aim to uncover details on how interviewees receive and make
  sense of sound, but also how any why they make it.
- *Web-based video calls*: Online delivery of interviews further enhances
  simplicity and cost effectiveness whilst also enabling the interviewee
  to participate in the more natural surroundings of their own home.
  Furthermore, it provides an ideal technological scenario for data capture. Interview audio is recorded alongside facial capture of both
  interviewee and interviewer by way of video. Using a desktop computer also enables the use of automated transcription software, producing a textual form with minimal research labour.

6 *Autoethnographic/soundwalk case study*: Interviews provide 'interviewee-
as-expert' knowledge but fail to capture 'researcher-as-expert' insights. It
is also very difficult to reliably establish a set of questions (or even topics
in some cases) prior to an interview without directly relevant preliminary
investigation. Autoethnography can provide this function, allowing the
researcher to synthesise topics and lines of questioning from their own experience, and also to apply their (*cough) expert knowledge to the subject
of study. It can also be used as a means of self-disclosure, passing the results
on to the interviewees as a means to establishing better rapport and trust.
For the purposes of this book, the chosen approach to autoethnography is
a series of bespoke soundscape recordings and soundwalks.

7 *Thematic content analysis*: This form of qualitative analysis bears the most
relevance to interview-generated data and autoethnographies. It also
possesses one of the strongest precedents and pedigrees in the field and
presents the possibility of conducting analysis either manually or automated by way of software.

8 *Technology*: Use of audio recording technology to facilitate repeat listening of captured everyday soundscapes will support more detailed and reliable analysis, but we must be careful to make a full account of the process, as means of capturing the sound is itself a component of sonic ethnography. Due to logistical and cost constraints, audio recordings will only be possible as part of the autoethnographic part of the study.

   — *Head-mounted binaural microphone*: Binaural microphones provide enhanced spatialisation of the soundscape, allowing the listener to experience the location of sound objects as they were experienced at the point of capture. Head-mounted capture provides the additional benefits of being easily portable and discrete, allowing the wearer to move freely and avoid drawing unwanted attention.
   — *Google Meets videoconferencing*: A light and efficient videoconferencing system that runs effective through a browser, thereby not requiring any software to be installed by the contributor and facilitating desktop, laptop, and mobile platforms. Recordings are automatically sent to *Google Drive* for easy cataloguing and review.

Chapter 2 summary: so, how did it all go then?

Apologies if this shift in tense is a little jarring. I considered leaving all of the retrospective until the end of the book before realising that I'd need to cross-reference a great chunk of material from this chapter to put my final thoughts in proper context. That, and I expected you'd rather know how it went now, whilst all the methodology detail is still fresh in your mind.

In practice, I found that many of the recommendations made by Hammersley and Atkinson (2007) could indeed be embedded into the interviews instinctively. Elements such as establishing rapport and avoiding potentially threatening or invasive questions proved straightforward enough, provided attention was regularly paid to the non-verbal communication of the contributor. Their vocal tone and facial expression often revealed their feelings on whether I had gauged matters correctly. Avoiding 'co-construction' was a little more challenging and it seemed to clash with the other recommendation of establishing rapport through self-disclosure. If a contributor raised a point on which I had a personal anecdote, that could (and often did) really help break the ice of the discussion. However, on several occasions, it felt as though an unintended consequence of doing this was that it did direct the next 10–15 minutes of the conversation, which became more of a dialogue than would have been desirable. Lastly, having a list of topics to broadly cover but not an explicit set of questions that had to be answered absolutely made the interviews themselves more relaxed and enjoyable. Less pressure to meet a set of targets encouraged more exploratory qualities within the discussions and meant, as the interviewer, I was free to simply listen to what

I was being told, without also needing to maintain concurrent attention on 'conversation quotas'.

In terms of the technology, well that was hit and miss. Online videoconferencing, namely Google Meets, proved perfectly reliable and easy to set up, and recordings were indeed very easy to catalogue, access for transcription, and review. The 'miss' came in the form of transcription software, though this was arguably a composite problem. Across all of the various solutions I tested, the speech recognition of my own voice was perfectly accurate, but I was using a pretty posh condenser microphone and I remained very close to it at all times. The audio hardware setup of my contributors was inconsistent. Whilst a few used comparable setups to myself, the vast majority used their phone or laptops' integrated microphone from which they were typically positioned at a sizeable distance. The speech-recognition software didn't stand a chance. This meant I had to manually transcribe *every single interview*. On the off chance, you're considering doing any interview-based research, and if you don't want to manually transcribe, I beseech you – conduct your discussions in person and get yourself *two* very good microphones. Thank me later. For the soundwalks I used a pair of Sennheiser Ambeo Smart headphones[7], wired into an Apple iPhone 12. The setup was very subtle and the fidelity of the binaural recording was impressively detailed, though at the time of writing the Ambeo headphones have been discontinued.

Finally, with regard to the recommendations put forward by Burns (2021), frequency analysis proved a useful means of establishing more potent themes and perspectives. The hierarchy of significance, particularly the features of absence, relevance, and critical incidents, proved efficient and not too labour intensive despite having over 120 hours' worth of video content to process. Lastly, assertion development actually proved to be less of a chore and more 'impossible to resist' as, in so many cases, the notions raised by many of the contributors compelled me to investigate further to see if what I had just been told had been explored within the academic literature. Of course, the best of these findings from the literature are distributed throughout the subsequent chapters, hopefully in complement to our contributors' insights.

So, have my feelings on qualitative research changed? Did this experience confirm my ignorance of, and bad attitude towards, this huge chunk of scientific methodology, or did it convert me? That, I'm not going to tell you yet. You can read the book to the end to find that out.

Notes

1 Link to Society for Ethnomusicology website: https://www.ethnomusicology. org/page/AboutEthnomusicol.
2 See the lyrics of Pulp's 1995 song *Common People* for a nice example of how simply being in the same place does not always equal being in situ.
3 Link to the British Library's soundscapes collection https://sounds.bl.uk/ Environment/Soundscapes (may require UK university institutional login to access).

4 Soundwalk in Central Park, New York City, USA: *Nomadic Ambience* – https://www.youtube.com/watch?v=kGJr1Nh-1CY&t=218s (accessed 07.02.2022).
5 Soundwalk in Lahore, Pakistan: *Keezi Walks* – https://www.youtube.com/watch?v=O1AQcFpPs04 (accessed 07.02.2022).
6 The 'ears on a stick': https://3diosound.com/products/omni-pro-binaural-microphone (accessed 17.02.2022).
7 Sennheiser Ambeo Smart headphones: https://en-uk.sennheiser.com/in-ear-headphones-3d-audio-ambeo-smart-headset (accessed 17.02.2022).

References

Adams, T. E., Ellis, C., & Jones, S. H. (2017). Autoethnography. *The International Encyclopedia of Communication Research Methods*, 1–11.

Adams, T. E., Jones, S. H. & Ellis, C. (2015). *Autoethnography*. New York: Oxford University Press.

Bauman, L. J., & Adair, E. G. (1992). The Use of Ethnographic Interviewing to Inform Questionnaire Construction. *Health Education Quarterly*, 19(1), 9–23.

Boehm, B., & Turner, R. (2003). Using Risk to Balance Agile and Plan-Driven Methods. *Computer*, 36(6), 57–66.

Brewer, J., (2000). *Ethnography*. London: McGraw-Hill Education.

Bull, M. (2018). *The Routledge Companion to Sound Studies*. New York: Routledge.

Bull, M., & Back, L. (Eds.). (2003). *The Auditory Culture Reader*. Oxford: Berg.

Burns, A. (2021). Analyzing ethnographic data. Ethnography Made Simple. Online article: https://cuny.manifoldapp.org/read/untitled-fefc096b-ef1c-4e20-9b1f-cce4e33d7bae/section/cc7fef6e-6dbb-4f16-b491-27ccb28d5f91 (accessed 13.02.2022).

Butler, T. (2007). Memoryscape: How Audio Walks can Deepen Our Sense of Place by Integrating Art, Oral History and Cultural Geography. *Geography Compass*, 1(3), 360–372.

Eckersley, R. M. (2016). Is the West Really the Best? Modernisation and the Psychosocial Dynamics of Human Progress and Development. *Oxford Development Studies*, 44(3), 349–365.

Ejnavarzala, H. (2019). Epistemology–Ontology Relations in Social Research: A Review. *Sociological Bulletin*, 68(1), 94–104.

Feld, S. (2015). Acoustemology. In D. Novak and M. Sakakeeny (eds.) *Keywords in Sound* (pp. 12–21). Durham and London: Duke University Press.

Garner, T., Grimshaw, M., & Nabi, D. A. (2010). A Preliminary Experiment to Assess the Fear Value of Preselected Sound Parameters in a Survival Horror Game. In *Proceedings of the 5th Audio Mostly Conference: A Conference on Interaction with Sound*. 9-15th September, Piteå, Sweden.

Glaser, B. G., Strauss, A. L., & Strutzel, E. (1968). The Discovery of Grounded Theory; Strategies for Qualitative Research. *Nursing Research*, 17(4), 364.

Greed, C. (1994). The Place of Ethnography in Planning: Or Is It 'Real Research'? *Planning Practice & Research*, 9(2), 119–127.

Grimshaw-Aagaard, M. N. (2018). What Is Sound Studies? In Bull, M. (ed.) *The Routledge Companion to Sound Studies* (pp. 31–44). New York: Routledge.

Hammersley, M., & Atkinson, P. (2007). *Ethnography: Principles in Practice*. New York: Routledge.

Harrison, K. (2012). Epistemologies of Applied Ethnomusicology. *Ethnomusicology*, 56(3), 505–529.

Hensen, B., Mackworth-Young, C. R. S., Simwinga, M., Abdelmagid, N.,... & Weiss, H. A. (2021). Remote Data Collection for Public Health Research in a COVID-19 Era: Ethical Implications, Challenges and Opportunities. *Health Policy and Planning*, 36(3), 360–368.

Hood, M. (1957). Training and Research Methods in Ethnomusicology. *Ethnomusicology*, 1(11), 2–8.

Howlett, M. (2021). Looking at the 'Field' Through a Zoom Lens: Methodological Reflections on Conducting Online Research During a Global Pandemic. *Qualitative Research*, 22(3), 387–402.

Humphreys, M., Brown, A. D., & Hatch, M. J. (2003). Is Ethnography Jazz? *Organization*, 10(1), 5–31.

Jääskeläinen, R. (2010). Think-Aloud Protocol. *Handbook of Translation Studies*, 1, 371–374.

Jeon, J. Y., Hwang, I. H., & Hong, J. Y. (2014). Soundscape Evaluation in a Catholic Cathedral and Buddhist Temple Precincts Through Social Surveys and Soundwalks. *The Journal of the Acoustical Society of America*, 135(4), 1863–1874.

Kahn, J. M., Rak, K. J., Kuza, C. C., Ashcraft, L. E., Barnato, A. E., Fleck, J. C.,... & Angus, D. C. (2019). Determinants of Intensive Care Unit Telemedicine Effectiveness. An Ethnographic Study. *American Journal of Respiratory and Critical Care Medicine*, 199(8), 970–979.

Kelman, A. Y. (2010). Rethinking the Soundscape: A Critical Genealogy of a Key Term in Sound Studies. *The Senses and Society*, 5(2), 212–234.

Kennedy, J., Holcombe-James, I., & Mannell, K. (2021). Access Denied: How Barriers to Participate on Zoom Impact on Research Opportunity. *M/C Journal*, 24(3), 1–11.

Kvale, S. (1994). *Interviews: An Introduction to Qualitative Research Interviewing*. Michigan: Sage Publications, Inc.

Liu, J., Kang, J., Behm, H., & Luo, T. (2014). Effects of Landscape on Soundscape Perception: Soundwalks in City Parks. *Landscape and Urban Planning*, 123, 30–40.

Maccoby, E. E., & Maccoby, N. (1954). The Interview: A Tool of Social Science. In G. Lindzey (Ed.) *Handbook of Social Psychology* (pp. 449–487). Cambridge, MA: Addison-Wesley.

Madden, R. (2017). *Being Ethnographic: A Guide to the Theory and Practice of Ethnography*. London: SAGE Publications.

Mara, A. F., Potts, L., & Bartocci, G. (2013). The Ethics of Agile Ethnography. In: Albers, M. & Gossett, K. (eds.). *Proceedings of the 31st ACM International Conference on Design of Communication* (pp. 101–106). New York: ACM.

Medelyan, A. (2021). Coding qualitative data: How to code qualitative research. Insights Thematic. Online article. https://getthematic.com/insights/coding-qualitative-data/ (accessed 13.02.2022).

Pinch, T., & Bijsterveld, K. (2004). Sound Studies: New Technologies and Music. *Social Studies of Science*, 34(5), 635–648.

Post, J. (2004). *Ethnomusicology: A Research and Information Guide*. London: Routledge.

Ramanathan, R., Paramasivam, S., & Hoon, T. B. (2020). Discursive Strategies and Speech Acts in Political Discourse of Najib and Modi. *Shanlax International Journal of Education*, 8(3), 34–44.

Rice, T. (1996). Toward a Mediation of Field Methods and Field Experience in Ethnomusicology. In G. Barz, G and T. Cooley (eds.) *Shadows in the Field: New Perspectives for Fieldwork in Ethnomusicology* (pp. 101–120). New York: Oxford University Press.

Rice, T. (2018). Ethnographies of Sound. In Bull, M. (ed.) *The Routledge Companion to Sound Studies* (pp. 239–248). New York: Routledge.

Saunders, M., Lewis, P., & Thornhill, A. (2009). *Research Methods for Business Students*. Harlow: Pearson Education.

Schafer, R. M. (1970). *The Book of Noise*. Wellington: Price Milburn.

Schuursma, A. B. (1992). *Ethnomusicology Research: A Select Annotated Bibliography* (Vol. 1). New York: Garland Publishing.

Spradley, J. P. (2016). *The Ethnographic Interview*. Long Grove: Waveland Press.

Sterne, J. (Ed.). (2012). *The Sound Studies Reader*. New York: Routledge.

Tickle, S. J. (2017). Ethnographic Research with Young People: Methods and Rapport. *Qualitative Research Journal*, 17(2), 66–76.

Valenzuela, D., & Shrivastava, P. (2002). *Interview as a Method for Qualitative Research*. Presentation given at the Southern Cross University and the Southern Cross Institute of Action Research (SCIAR).

Wong, D. (2014). Sound, Silence, Music: Power. *Ethnomusicology*, 58(2), 347–353.

Wutich, A., & Brewis, A. (2019). Data Collection in Cross-Cultural Ethnographic Research. *Field Methods*, 31(2), 181–189.

Young, J. C., Rose, D. C., Mumby, H. S., Benitez-Capistros, F., Derrick, C. J., Finch, T.,… & Mukherjee, N. (2018). A Methodological Guide to Using and Reporting on Interviews in Conservation Science Research. *Methods in Ecology and Evolution*, 9(1), 10–19.

3 Sound and meaning

South Pier, Southsea. 4th of September 2021. Late afternoon

I am walking towards the southern pier. The soundscape changes dramatically. The first sound to cut through the air is the hiss of hydraulics. I cannot yet see the funfair, but this sound immediately stimulates the imagery of what I'm about to see. The looping funfair music fades into the scene. The plodding, two-step musical phrases burrow into my head even as they crash into one another. What is noticeable is that these sounds are effectively the full extent of what I can hear. Human voices are rare, partially drowned out by the music but mainly just absent. The funfair is not attracting many visitors at this time. The soundscape evokes a slight melancholic effect, as if the fair is reaching out to people with its music, inviting them to come and play, but the people are not engaging. Yet the sounds of the fair continue, seemingly in vain. Like a film playing in an empty theatre.

Well, here we are then. I'm not sure I can overstate the size of the exhalation that followed clicking 'save' on the draft transcript for the final contributor. Of course, any feeling of relief was pitifully short-lived as now began the process of trying to make some overarching sense of the data that now sat in front of me. I can't say I hadn't been warned by my past-self about the challenge that lay ahead, namely the process of classifying and collating the hundreds of pages of transcriptions into a meaningful structure. If I ever needed a reason to back away slowly from my newfound relationship with qualitative data, this one was pretty compelling. As discussed in the previous chapter, the presentation of the questions, from their precise wording to their order of appearance, even to whether they were addressed or not, was all flexible and responsive to the interviewee. That of course meant no two interviews followed exactly the same structure. Many of the answers provided by the contributors were also highly personal. Even if asked the exact same questions, responses would have still been difficult to directly compare. Broadly classifying the

DOI: 10.4324/9781003178705-4

various passages of each transcript in relation to specific questions also proved decidedly difficult, as many contributors would provide an answer to certain questions that, upon review, felt much more relevant to an altogether different question. In many instances, they would respond with initial information that was relevant to the question asked, but then proceed to explore divergent topics. This may be less of a balanced retrospective and more of a ham-fisted attempt to lower your expectations. To counteract this a little, it bears repeating that the value of qualitative interview data is precisely its individuality, its richness, and its reflection of personal experience, interpretation, and meaning. So, did the information collected deliver?

Immediate reactions

The question *"what does sound mean to you?"* quickly became established as the opening topic of the interview. It was the least susceptible to variations. Even my little preambles to it became decidedly uniformed from one interview to the next. With this question, I was hoping to receive a genuinely diverse range of answers that cut to the heart of the interview and, through the openness of its phrasing, would invite more personal insights and less 'stock' answers. The immediate, instinctive responses were, for many interviewees, an exclamation at the openness of the question and also the breadth of scope in their potential answer.

> I think that this is a very difficult question.
>
> Anonymous: China, Hong Kong

> That's a big one, isn't it?
>
> Stafford: UK

> That is a huge question!
>
> Anonymous: Italy

These were more than fair observations. In prefacing this first question, I found myself increasingly empathetic in the challenge of trying to describe the meaning of sound without any preparation. Although I did worry that any forewarning would itself be anxiety-inducing, I couldn't help but preface with something along the lines of: *"Apologies in advance, this is a horribly broad and open question"*. Fortunately, although the initial responses of many were a blank stare, with possibly a wee bit of nervous laughter, every contributor composed themselves and eloquently pressed on. In many instances, their responses foreshadowed the themes and ideas that would feature in their answers to later questions and yielded some distinct and emotive viewpoints.

> Sound gives me a sense of everything.
>
> Meni: Mexico

Sound is about our history of experiences. From some of the sounds that I hear, I am reminded of my journey throughout life, back as far as I can remember and up until today. Memories of being with family members or friends, and of certain activities like being at the beach or building snowmen. All of those memories, some good and some bad, are wrapped up in sounds and voices. Sound is an input, connecting me to the outside world.

Mahmood: Iran, UK

I think meaning in sound can be very different for different people. So, whether it's absolute silence or extremely loud construction noise you hear in a given moment, sound has the power to create different emotional implications and experiences for us.

Can: Turkey, China

Sound means life. As long as I can hear different details of sound, I'm alive. Sound gives me dimensions to what I see. Sound gives me emotional guidelines on how to read what I see. Connecting to sound is both conscious and unconscious. When I hear something, it can evoke memories from throughout my whole life.

Heikki: Finland

Sound means a lot. It means space, movement, colour, and position. It connects with story and feelings. For me, sound is space and colour most prominently. I visualise sound in terms of shapes and colour, then use that to position them in my mind, in an imaginary space.

Nicolas: Iceland, France

I find that certain sounds become more embedded in my memory than some other senses.

Ludo: Italy, Egypt, UK

I think that sounds can affect us emotionally, physically, consciously, and unconsciously. They affect us through memories, they can trigger remembering of elements from the past. This can be both musical elements and non-musical types of sound. So, if a particular piece of music was playing during a personally-important road trip for example, where we were with our family and experiencing happy moments; hearing that same music at a later time, we would recall those memories and feel happy, just as we did in the actual moment.

Andreina: Venezuela

Sound has properties that communicate information, although that information may not always be meaningful to us. In itself, sound doesn't mean much at all, but what we can interpret, what we can feel, and what we can do in response to a sound is very meaningful but also highly personal.

Anonymous: Senegal, Togo

I think sound takes on different meanings over time, especially based on where you live, your age, as your tastes change, and so on.

Alyson: UK

In a few short snippets of responses, sound means: a unique experience, responsive to and reflective of our individuality; a link to the past, a way of accessing our personal history; a gestalt experience, much greater than the sum of its parts; a constancy of experience, immersive and ever-present; and a passion, explored through hobbies and potentially through a wide range of professional practices. Its meaning is also deeply personalised, with our experiences, preferences, understandings, expectations, and more, all culminating in a unique perceptual processor, from which our own sense of meaning emerges.

Emergent themes

Objective aspects of sound

Respondents varied quite dramatically in both the scope and focus of their answers. Some would respond very succinctly. Others would elaborate on their ideas for quite some time, largely thinking out loud, and often taking brief moments to pause and consider their own commentary. Here, a personal favourite:

Sound is what I can hear with my ear.

Anonymous: Italy

There's an element of beauty in the succinctness of this statement. Whilst many could be quick to insist that sound is much more than simply what we hear with our ears, few would suggest this statement to be false. What we hear is indeed a foundational component of the meaning of sound. Indeed, many further responses to this question would commence with statements akin to the acoustic definition of sound.

Sound is the vibration that is propagated through a transmission medium.

Marc: Mexico, China

Vibrations that travel through a medium and give a certain sensation to the human senses.

Anonymous: Indonesia

It's a medium of transmission. It comes from a source to a living creature that is able to listen.

Anonymous: Peru

There's a little caveat to this observation. The individuals who focussed, or
at least prioritised a more acoustic meaning of sound largely (though not
entirely) came from an audio engineering background. The influence of
working professionally with sound upon its perceived meaning is certainly
something that we shall be returning to but, for now, it is worth noting
that simply working with sound was unlikely to mean that the interviewee
would automatically champion the acoustic definition, if acknowledge it at
all. Furthermore, many who began with this focus quickly moved on to more
perceptual, creative, and technological elaborations.

Science tells us that sound is energy, that it moves through the air and the
water. It enters our ears, reaches the auditory cortex and then we listen;
we interpret the sound through listening. Obviously, I agree with that,
but I also feel that sound is an invisible force that triggers and elicits different kinds of emotions and sensations in our bodies.

Andreina: Venezuela

Physical sound is energy propagating through a medium. It's a technical
way of seeing it, but it also opens up a lot of possibilities for what sound
can be. Like EMF recordings, which are Electro Magnetic energy waves
emanating from electric devices where the energy has been 'converted'
into a humanly audible range. With digital sound there's no more limits
to what sound can be. It's possible to convert audio into a photograph,
edit the image then return it to audio, creating a whole new sound.

Rasmus: Denmark, Sweden

As contributors from professional sound design, engineering, or research
backgrounds proceeded to speak more subjectively, they did so in most cases
with quite a marked degree of emotional expressiveness and passion. This
dispelled the idea that analytical and affective relationships with sound cannot
co-exist but rather reinforced the notion that as one grows, so does the other.

Meaning through definition

In the first chapter of this book, we pondered the question *'what is sound?'*. We
first looked at the acoustic definition that identifies sound as a soundwave,
before considering sound as an object, an event, an object, *and* an event and
sound as a holistic phenomenon. For many, sound and meaning is a relational
matter that concerns the connection of sound to listener. Indeed, my wording
of that opening question *'what does sound mean to you?'* invites answers that

consider sound as a discrete entity, but one in relation to the person providing the answer.

Prior to the interviews, all of our contributors kindly completed an initial questionnaire that allowed me to set up the videoconference but also collect some additional data that could be more easily collated and compared between contributors. One of these additional questions presented the contributors with a simple hypothetical: "*You hear a dog barking in a park. Someone asks you what you hear. Which of these answers do you feel is the most accurate?*". Drawn from the relevant academic research on the subject (see Chapter 1), a total of eight possible options were available:

- *Sound is a wave.* I hear a soundwave, produced by a dog as it barks
- *Sound is an object.* I hear a dog
- *Sound is an event.* I hear barking
- *Sound is the property of an object.* I hear the bark of a dog
- *Sound is a perception.* What I hear evokes the idea of a dog and that it is barking
- *Sound is an object or event in relative space.* I hear a dog barking to my right
- *Sound is a phenomenon.* I hear a dog barking, to my right, in a park
- Other...

The responses to this question are visualised in the two figures below. As the respondents kindly identified their professional background as either 'non-sound professional' (Figure 3.1) or 'sound professional' (Figure 3.2), it was possible to split the data between these two groups to review the results separately.

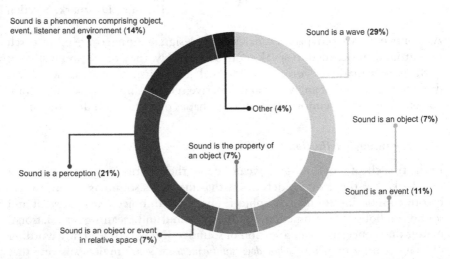

WHAT IS SOUND?

Sound is a phenomenon comprising object, event, listener and environment (**14%**)

Sound is a wave (**29%**)

Other (**4%**)

Sound is an object (**7%**)

Sound is the property of an object (**7%**)

Sound is a perception (**21%**)

Sound is an event (**11%**)

Sound is an object or event in relative space (**7%**)

Figure 3.1 Multiple-choice results to the question: *what is sound?* Non-sound professionals

WHAT IS SOUND?

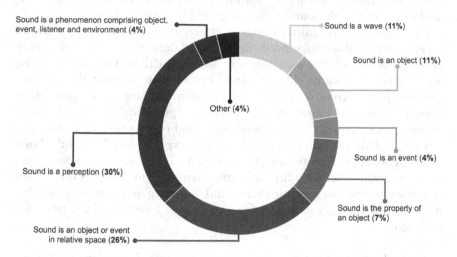

Figure 3.2 Results to the question: *what is sound?* Respondents were all sound professionals

There is no statistical analysis here. No *t*-tests, no checks for if $p < .05$, and *definitely* no regression analysis. Just the most surface level of descriptive analysis as a means of raising questions, not trying to answer them. Despite being superficial enough to offend a good few thousand PhD students (they tend to get annoyed when someone claiming a scientific background gets away without doing 'proper statistics'), there are a few interesting things we can observe. First, both groups acknowledged every definition, with no single option being left completely unchecked. Second, again in both groups, two particular definitions took notably greater shares, with *sound is a perception* being a popular choice for both sound professionals and non-sound professionals.

There were also a couple of surprises. The first was that a much greater percentage of non-sound professionals identified sound as a soundwave. My initial assumption was that those working with sound would hold a more objective, technical view of sound, akin to the acoustic theory and favouring the soundwave as the physical manifestation of sound. My second surprise was that the most popular definition for the sound professional group was *sound is a perception*. Whether there is any truth to this I honestly cannot say (the PhDs and statisticians would come for my head), but my hunch is that these surprises are due to the distribution of specific sound professional roles. Most of our contributors identified as sound designers, with only a handful of engineers. This pool also largely featured contributors producing sounds for film, television, theatre, or digital games. This would (hypothetically) explain their favour towards *sound is an object or event in relative space*, with 3D/

spatial sound being a core aspect of sound design for these industries. It would also explain their preference for *sound is a perception*, with audience reaction to (and interaction with) sounds also being vital in their work.

Returning to that initial question concerning the meaning of sound, various contributors referenced one or more definitions. From the interview responses below we can see how some first thoughts were to define sound through a more objective lens, but then to connect this to more personal perspectives on matters of emotion, art, music, or perception. Personal history and memory are described as providing a unique listening experience, meaning a single sound presented to two people, or one person at two different points in time, could be experienced differently. Some respondents appear more object-oriented whilst others are more drawn to a composite understanding of sound, with the nuances in the various responses heavily suggesting that sound meaning is messy and flexible, with the listener's perspective subject to change based on place, time, and situation.

> I think of sound as a combination rather than a single element. When I am walking in the woods or the mountains, the sound is a compilation of the wind, the leaves, the sea breeze, the birds, but also the place of my walking.
>
> Anonymous: China, Hong Kong

> I often think of sound in scientific terms, so simply as a vibration, but vibration means something that is actually touching you, so you can feel it. This means that the hearing-impaired can also experience sound, but in a very different way. No two people will experience sound in exactly the same way, or attribute exactly the same meaning to that sound.
>
> Daniel: Costa Rica

> For me sound is an object. It is something that we hear and interpret.
>
> Anonymous: Senegal, Togo

> A sound is created by an object, but it's interpreted by us. We can attribute certain things to a sound, and we can imbue it with meaning.
>
> Miles: UK

> Sound has its scientific definition, but it's also metaphorical. I may hear a dog, but the way the dog is barking may have a particular meaning to me; not only based on what I am hearing but also what I have experienced of that in the past. It's always a story and not just of the sound, but of how I interpret that sound. I think that's what makes sound an artform.
>
> Arturo: Costa Rica

It depends on the context. If I was thinking academically, I would be thinking about sound in technical terms, like sound as a wave. I would also think about sound from an architectural perspective because my training in architecture involved courses in sound design. So, we would think about acoustic design, the use of acoustic panels, and different roof heights for example. But I think, more generally, sound is noise. Sound is music.

Dana: Jordan

Perceptual responses

Immersion and persistence in sound

In several interviews, the above notion of sound being a persistent presence was raised as a key feature of its meaning. Well observed was the basic tenet that we have eyelids to shut out incoming visual stimuli but no equivalent for sound. This was extended in some instances, with several contributors beginning to touch on what they felt were the immersive and permeating qualities of sound.

Sound for me has been life-changing. I remember, a friend once told me that the ears don't blink. Even when our eyes are shut, sound remains a constant sense where information can reach us. People in comas have been found to have heard and remembered certain sounds. We are washed in sound every day, twenty-four seven. Sound has always been there, even in my early childhood.

Stafford: UK

Sound is something that we cannot switch off. We can close our eyes to try and keep the outer world away from us and yet we cannot do the same with sound.

Ludo: Italy, Egypt, UK

Sound is an always-surrounding, mood-defying penetration of senses.

Anonymous: Germany

Sound for me is more than half of everything, in the sense that I'm constantly listening to something. I have tinnitus, so silence really doesn't exist, and I don't have a single memory that doesn't have sound. Sound is everywhere. Even if you cannot hear, you can feel the vibrations.

Angélica: Spain

The notion of sound experience being largely subconscious and not something that we are typically aware of was also a regular observation. This raised a question of the extent to which being persistently immersed in sound and our largely pre-attentive relationships with it are connected.

> Sound is everywhere, we just don't always perceive it. The refrigerator can be running in the background constantly, but we don't notice it until it goes off. You notice that there was a sound now that it's absent.
>
> Meni: Mexico

> Sound is with us from the very first moment that we are born. It's built into our subconscious.
>
> Mehrdad: Iran. UK

> Sound has a subconscious effect. It is part of our culture and part of our environment, and that has an effect on how you respond to it, whether you're aware of it or not.
>
> Nicolas: Iceland, France

The listener

The importance of the listener is asserted across multiple definitions of sound. Without a listener, some believe there simply is no sound, whilst others would suggest that sound remains but without a listener to receive and make sense of it, sound has no meaning. Throughout the interviews, the term 'listening' almost uninformedly described acts of purposeful engagement with sound.

> To me, listening is the usage of my hearing on purpose - whether it be a music album or talking to a friend. It's something that needs focus.
>
> Anonymous: Germany

> Listening is the ability to be present and give attention to sound.
>
> Anonymous: Indonesia

> Listening means carefully hearing sounds to try and understand them for a particular purpose.
>
> Marc: Mexico, China

Both musical and non-musical sound featured regularly as the focus of contributors' examples of listening. It is worth noting that few contributors used the term to describe engagement with both music and non-musical sounds, rather one or the other. It also became quite clear that most contributors from a non-sound professional background associated listening with musical sounds, whilst sound designers and engineers displayed a definite preference towards non-musical sounds, most predominantly the sounds of human voices, their immediate environment, or experiences in nature. For the contributors who considered the question of listening more in terms of definition, responses were mostly consistent and reflective of the observation made by Crawford (2009), that listening is broadly perceived across society as being akin to the act of paying attention. Across the responses, there was a clear accord on the act of listening comprising elements such as attention, focus, and intent. In terms of variation, some responses emphasised more analytical functions of listening, whilst others underscored more social and communication uses:

> it's a bit difficult to say what listening is because there are at least two ways of listening. Active listening is where you try to fully focus on what you're listening to. It can be in the studio, on headphones, or just when you're out taking a walk. How does the wind sound when you move at different speeds? What is the texture of the ground beneath you? How do sounds reflect off the environment around you? In passive listening you don't really engage with the soundscape.
>
> Rasmus: Denmark, Sweden

> Knowing exactly what to expect from sound, being able to identify its origin, its source and its mechanism for being produced can help you to anticipate things before other people. I remember, I was in the Marriot Hotel in Islamabad. We were on the sixth floor in my colleague's room, and I remember looking up from my laptop and saying: "there's an earthquake coming". He paused and listened for a few moments then dismissed what I had said. Five or six seconds later and I was turning white. I repeated my previous statement. I did this three times, each with that pause in between. At this stage I couldn't stay. I walked out and my colleague followed. We stepped through the door that the entire building started 'dancing'.
>
> Hatim: Pakistan

> For me, listening is more about understanding, sharing and respect. Catching the message.
>
> Si Qiao: China, UK

Active listening as a term goes back at least as far as 1957 (republished most recently in 2021) and the work of American Psychologists Carl Rogers and Richard Farson. Whilst sound is a core element within most concepts on active listening, the term appears most commonly in social communication. Rogers and Farson highlight requirements for active listening that encompass both reception and response. Reception needs to understand a message in terms of both content (non-decorative language that is less subject to interpretation) and affect (how the content is expressed in terms of vocal expression and decorative language). The response should acknowledge both the content in context of the affect and the affect in context of the content. It is worth noting however that this definition of active listening appears prominently in scientific literature when searching for the term exclusively. If you were to type "active passive listening" into your search engine[1], the results would reflect Rasmus' quotation more closely. Here, the results do highlight the interactive elements proposed by Rogers and Farson, just not limited to the function of social communication. It is fair to say, however, that there is more than a little inconsistency in the fine print of how the 'engagement' element of active listening is defined. Fung and Gromko (2001), for example, compared the effects of active and passive listening upon listening-enjoyment and ability to correctly identify rhythm and tempo. Here, active engagement described the act of spontaneously moving in response to the music. In contrast, Remijn and Kojima (2010) used the term 'active-response listening' to describe a group that was specifically directed to judge a rhythmic element of a sound, as opposed to a passive listening group that was instructed simply to "listen to the sound stimuli and randomly press one of two buttons after the end of each" (p.3). What does appear to be consistent, both here and in various other studies (Palmer et al. 2007; Górska et al. 2018), is that active listening will involve some form of task for the listener. This task may be self-prescribed or self-directed, or it may be imposed upon them, but there must be a sound element that the listener needs to interact with (be it cognitively, emotionally, physically, or all three) as a means of completing the task.

Functions and values

> Sound functions as a fundamental part of everyday life. Sound is for localisation, for identifying danger. But then there is sound for entertainment and enjoyment. This may involve music or sound design. If you play a video game or go to the theatre, you experience sound. As designers, we use sound to generate emotions and to tell or support a story.
>
> Daniel: Costa Rica

The affordances of sound

Continuing from matters of listening, particularly active listening, what sound fundamentally enables us to do within the world is arguably a natural follow-on. Affordance is a term that dates back to 1966 and the work of James Gibson and Leonard Carmichael, though the notion behind the term is arguably one that has been present as far back as the time of Aristotle (Smith 2000). For Gibson and Carmichael, affordance was not simply about what something can do, or by extension, what it can do for the listener. Instead, how something supported effective interaction within a person's environment was at the core of affordance. So, what can sound enable us to do within our environment? Returning briefly to our questionnaire results, the affordances of sound were also explored briefly by our contributors in a simple format that would hopefully produce lightweight but easily digestible results.

The question itself was presented as a three-point multiple-choice grid where contributors were asked to consider seven affordances of sound and select their perceived value between: (a) great importance; (b) moderate importance; and (c) no importance. As with the earlier question on defining sound, the results were split between contributors who were *not* professional sound designers, producers, or engineers (Figure 3.3 above), and those who were (Figure 3.4 overleaf), as a way of also uncovering any evidence to suggest that working with sound professionally could have an impact on a person's attitude towards it.

Whilst the interviews prioritise richness of information, hopefully, these figures provide an accessible, at-a-glance overview of how the contributors felt sound helps them in their daily lives. Undoubtedly, there is a little participant bias to account for, particularly acquiescence bias, drawing respondents in the direction of the more positive options. That issue aside, there arguably remains a clear skew towards perceiving sound to be of genuine value in several different ways. As a means of reading the emotional attributes of a given situation and as an evoking stimulus of memories and ideas, the affordance of sound is rated markedly high, both for those who work professionally with sound and those that do not. It was interesting to see that the affordances that were rated more towards moderate (localisation, navigation, and instruction) were arguably the more routine and continuous uses of sound. This raises the

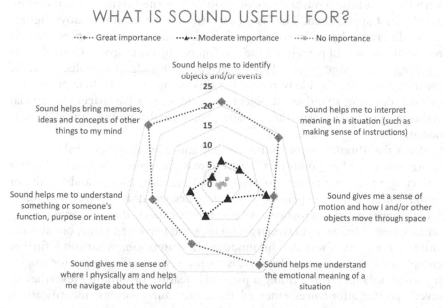

Figure 3.3 Results to the question: *what is sound useful for?* Respondents were not sound professionals

IN THE CONTEXT OF YOUR PROFESSION, WHAT IS SOUND USEFUL FOR?

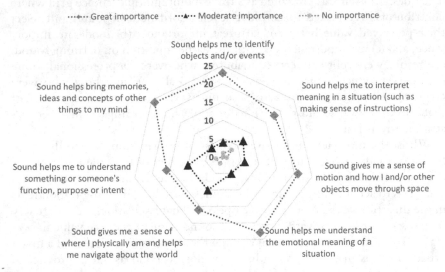

Figure 3.4 Results to the question: *what is sound useful for?* Respondents were all sound professionals

possibility that, ironically, the most vital affordances of sound are associated with the most innocuous and subconsciously governed actions, and therefore are the least appreciated. The last observation I take from the above figures is the difference between sound professionals and non-sound professionals on the affordance of perceiving the motion of objects in space. Contributors working with sound were a little more likely to judge this value to be of great importance, the likely reason for this being (as with their preference towards defining sound as an object or event in relative space) that spatial audio is a prominent element of many professional sound design and engineering roles.

What the affordances of sound can be, and why they have value to our contributors, are key points that feature across numerous responses. Whether reflecting upon connections to professional practice, friends and family, or fundamental sensory experience, sound is described as offering meaningful and appreciated affordances. As you might expect, there is some overlap within these discussions between affordance and perceived value, but also the emotional aspects of sound. This muddies the waters somewhat and is further complicated by most contributors discussing several affordances (or implying an affordance by describing a particular feature of sound). Nevertheless, seven specific affordances emerged from our conversations: identification, localisation, navigation, interaction, threat management, communication, and self-care.

Sound gives you a sense of where you stand in space. In terms of film, sounds are wonderful for telling a story. Sound is an ingredient in the recipe of poetry and add to beautiful imagery.

Nicolas: Iceland, France

China is a country with so many different cultures, religions, and behaviours. Living in big cities, sound is a way of knowing where I am, what is happening, and what is around me.

Anonymous: China, Hong Kong

Sound helps me to move through the world. Because it is invisible, not many people actually pay attention to it. Sound helps us to survive as a species. Going back thousands of years, we may hear the sound of a wolf and we would run and hide. Today, we would hear a fire alarm, and again we don't think, we just react.

Andreina: Venezuela

As Gaver (1991) points out, affordances can be multi-layered when relevant to more complex actions. They may present themselves in sequence, where acting on one affordance reveals a new one, or as 'nested affordances' that indicate smaller, component 'sub-actions' within a larger, more high-level action. One good example of this is described below:

It's almost like every sound is a communication that can tell us something. If I hear a car outside, that has the potential to tell me roughly how large the vehicle is, whether it is moving and if so, what its velocity is. It can also tell me its position and what direction it is moving in, all without me actually seeing it. I also think that motion is a key element in sound. It's interesting how motion is linked to the ear as well; that we can feel the differences in pressure as we move around, all through the ear. I think the experience of sound and motion is highly linked.

Miles: UK

Here, we have a description of physical properties as the nested affordance. As Miles explains, the sound of a car can provide us with multiple information points relevant to its physical properties. Attention paid to such properties, whether conscious or subconscious, can feed into higher-level affordances of identification (*"that sound is a car"*), localisation (*"I am walking in the path of a moving car"*), and navigation (*"I should probably move away to avoid getting run over!"*).

Subsequent commentaries that explored the meaning of sound from a functional perspective yielded a wide array of further perspectives.

Sound is a means of communication. It's a means of sharing information, both consciously and subconsciously. What you're trying to communicate and how you use sound will change, but when you use sound, you inevitably communicate something, even if you don't intend to. A dog might not intend to communicate information to you when it barks, but that doesn't mean it isn't.

Borneo: UK

I think that sound gives me an understanding of my environment; whether it is calm or whether there is turmoil, or fun. It suggests how I should feel and tells me how others are feeling. Should I be happy? Should I be worried? Are my neighbours fighting?

Anonymous: Kenya

Sound can alert us when something dangerous is happening. It can change our mood and enables us to communicate and interact with other people. Sound gives you a kind of feedback loop when you're doing something, say pressing a key at a computer. That clicking sound you get from the keyboard is feeding back the idea that you've done something. Old CRT televisions mechanically gave you sound feedback when you switched them on our off. They generated a magnetic field, and you would hear the charge building up and discharging. You don't get that in modern televisions.

Mehrdad: Iran, UK

Sound is a mechanism by which we convey information, be that your alarm clock informing you that it's time to wake up, or speech from your work colleagues. There is also naturally occurring sound that provides us with information about our surroundings. More than anything else, sound is an information source relevant to our environment and to our activities within that environment.

Anonymous: UK

I can use sound to recognise things. Sound can inform you so much about different activities; so much information without seeing it. I do need to be aware of the sound around me. It's part of how I work, how I function.

Smita: India

Sound helps me to relax. It can change my mood and my attitude. It helps me capture other people's moods. Sound is often a message. It helps me to understand people and what they are saying. It's an essential element to help me to sense things. Sound identifies the object and the scenario that is happening. It asks you to do the next action or reach the next understanding. You don't necessarily have to do everything sound tells you to do, but you need to evaluate the sound to make that decision. We use sound to identify objects in the world. Conversely, an unknown sound evokes curiosity. If a sound is unfamiliar, you can wonder what it might be, driving you to explore.

Si Qiao: China, UK

Without sound I would always feel that I am not as well equipped to act in the world to the fullest extent. If I am speaking with someone, how I respond to them will matter greatly on the emotion I am receiving from their voice. You might start a conversation with certain expectations, but then you can adapt, based on sound cues, to help you interact with people.

Hatim: Pakistan

Some people might not consider sound to be very important or just take it for granted. I find that extremely interesting because we as humans are always designing sound, and mixing, for our life: playing music while doing certain activities or creating playlists for situations, selecting ring tones, even specific ones for certain people, opening and closing windows or doors to change the volume of certain sounds, amongst many others. We constantly make decisions that ultimately shape the soundtrack of our life. It has to do with instinctive, or pure human behaviour.

Daniel: Costa Rica

My radio is on 24 hours a day, even if I'm out I will leave it on for my dog. I find it very difficult to be at home with absolutely no noise. I struggle to work without the extra background noise of the radio or the television. When I lived in London I would often work at home during the evenings, at which point I always had to have the television on so that there were still voices in the room and it wasn't just me, alone in my home.

Alyson: UK

I used to sing in a very big choir, so for me, sound gave me community. It helped me engage with other people, express messages, information, or emotions. Sound can also mean danger.

Lucia: Spain

As I learned in my very first year as a researcher, I do enjoy a good conceptual map. I enjoy staring at them. I enjoy feeling as though I'm dragging disparate pieces together with my bare hands, fighting against some sort of magnetic resistance that desires to keep the elements apart, lest they dare make any sense to anyone. In that spirit, I do hope that this little diagram (Figure 3.5 below) is helpful.

Each element within the map was added, based on the above interview responses. Sticking with Gaver's (1991) notion of nested affordance, the left-most elements represent the most fundamental. You could alternatively

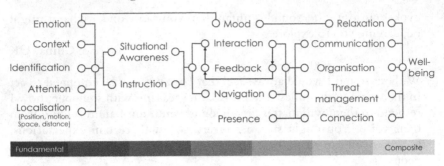

Figure 3.5 Conceptually mapping the affordances of sound based on interview responses

describe them as the most direct or immediate attentional affordances of sound. They also align quite well with Tuuri and Eerola's (2012) model of listening modes (see Chapter 1). Here, our fundamental affordances resemble their 'denotative modes', which include identification, instruction, and emotion. Further affordances of sound, identified by the contributors, fit rather cleanly into our own model as composites of the more fundamental affordances. Sounds enable us to prioritise elements of the environment by 'pulling' our attention towards them. It offers us the ability to identify the objects around us, to assess their relative position and motion in space (and, consequently, also our own), or to extract any possible instruction, affective content, and basic contextual information.

Through the fundamental layer of sound affordances, we are granted situational awareness, an understanding of our environment and our place within, and an instructive sense of how we should act within these surroundings. This affordance offers us further agency to move successfully around our environment and to interact with the elements within it. Here, sound provides the additional (and pivotal) affordance of immediate feedback that can confirm or correct our actions. It additionally provides us with a sense of presence. This can be about making *you* feel present within a given environment, but it can also be about feeling other entities (including people) as being present in the world with you. Subsequently, we have yet another layer of affordance as navigation, interaction, presence, and feedback feed forward into interpersonal communication and more complex social interaction, but also to enable us to assess physical and emotional threats. This gifts us the affordance of safety. We are also able to build structures of behaviours, navigations, and interactions through sound, meaning that sound helps us to organise ourselves within an environment through identification of, and adaptation to, external organisations. Lastly, it gives us the affordance of connection. It ties us to the world, making us feel a part of something much bigger than ourselves.

This multi-layered nest of affordances culminates with the mood-regulating and -relaxing potential of sound presenting us with self-care, the affordance

atop the pyramid. This is not to say that our respondents successfully identified every possible affordance of sound. However, the great range of affordances discussed, without prompts from myself, was certainly impressive.

The value of sound

The above responses demonstrated an awareness amongst contributors in
terms of the function of sound, how we may use it and what capabilities it
can afford us. A closely connected theme many contributors pointed to was
the value of sound. I have chosen to separate this from function despite the
overlap, because several respondents did not expressly discuss anything that
could be identified as function or affordance, yet absolutely expressed considerable gratitude for having the ability to hear. In some cases, the gratitude
was followed up with concerns over the future hearing loss and the great
need to protect our hearing. Either way, providing a full analytical account
of sound's affordances was not necessary for a person to express a profound
appreciation for sound.

> I think that sound is very important. It features in many different con-
> texts. In advance of this interview, I was thinking about what life would
> be like if I couldn't hear anything. At this moment, but also since I was
> very young, music has always been very important, both for me and my
> family.
>
> Anonymous: Malaysia

> I'm also conscious that our human capability to perceive sound can de-
> grade with time and it can be very difficult to preserve it. My father be-
> gan showing signs of losing his hearing in his mid-60s and now that he is
> in his 70s, for example, he will watch the television with the volume up
> to extremely loud levels. I worried about this even when I was younger,
> and I have always taken steps to try and protect my hearing. Out in the
> street, if there are very intense noises, especially the sound of a jackham-
> mer or a passing ambulance, I will actively cover my ears or will wear ear
> plugs if I'm in places or situations where I know it will be very loud. Even
> if it is only a very brief sound, I know that, cumulatively, sounds of such
> intensity can be damaging to my hearing. One minute of loud sound isn't
> a problem, but one minute every day for a year can be.
>
> Ludo: Italy, Egypt, UK

> From a very basic level, when I would meet someone who can't hear,
> who is deaf or has a hearing impairment, I do think "I'm really lucky
> aren't I?". That I have the ability to hear sound is, to me, something for
> which I feel privileged.
>
> Anonymous: Norway

I think, for people who have all of their five senses, individual sensory inputs are something they won't really consider because they'll experience all input as merged within the information sphere in which they live. I think if someone is deprived of one or more senses, it becomes a lot more pertinent. By not being able to experience sound, it will have a significant personal meaning.

Anonymous: UK

A few weeks before I was due to submit this manuscript to the publisher, I was not taking the best care of my physical or psychological health. My caffeine consumption was, ahem, excessive. My sleep patterns were erratic, and my exercise routine was non-existent. One night, during this period, I took myself off to bed. As my head hit the pillow, I noticed a distinct ringing in my ears. The tone was similar to that of a prayer bowl. The sound was still present the next morning and the following night, by which point I'd started to feel genuinely concerned. Fortunately for me, the ringing subsided over the next two days, but it gave me an education on the feeling of tinnitus and its possible impact on my daily life.

Though a common problem, tinnitus is generally agreed to be a poorly understood condition. We know that it is more prevalent in men than in women and that it is strongly correlated to hearing impairment, particularly in the age groups under 75 years (Lockwood et al. 2002). Research and medical understanding broadly agrees that there is no pharmacological treatment for tinnitus (Bauer 2018). Individuals suffering from the condition must therefore find strategies to manage it. A review by Lockwood and colleagues (2002) consolidates much of the relevant information on tinnitus into an evaluation framework that helpfully illustrates the most salient points on the condition. Here, the assessment commences with an examination of the patient's daily sonic environment and the effect this has on their quality of life. It will also consider family history of hearing loss and any current or historical usage of ototoxic substances (drugs with known connections to the auditory and vestibular systems). The assessment then seeks to differentiate objective and subjective tinnitus. Objective tinnitus describes circumstances in which we are hearing sounds with a physical correlate, from within the body. Causes can include vascular anomalies, atypical muscular activity, and various other spontaneous otoacoustic emissions (sounds produced by the inner ear itself). Subjective tinnitus is much more common. It presents no physical correlate and its cause can be extremely difficult to pinpoint. This is largely immaterial, however, as the advice broadly remains the same, with patients instructed to avoid exposure to loud noise wherever possible and, if the effect of the condition is great, to explore the potential of anxiety, depression, and sleep disorder therapies, alongside management techniques such as cognitive behavioural therapy and sound masking (e.g. white noise).

In my research responding to my own experience, I found that quite a few myths on tinnitus persist, despite being debunked by research. The most prominent of which is the belief that caffeine causes, or at least aggravates, the condition. I must admit, in my experience, I did instinctively assume a connection between the ringing in my ears and the greatly increased quantity of caffeine I was consuming. For those of you worried it's a choice between ringing ears and life without your morning flat white, or three, this link has not only been shown to be without evidence, but that caffeine actually presents a negative correlation to tinnitus diagnoses. Abstaining from it can actually worsen symptoms in some observed cases (Claire et al. 2010).

There is a recurrent theme within this book, that many of us can take our hearing for granted and may be unwittingly subjecting ourselves to risk factors for hearing loss. Scientific research has considered such risk factors for many years, with prominent factors including cardiovascular health, particularly issues connected to smoking and increased alcohol consumption (see Lin et al. 2012). Research has also pointed to ethnic variation in hearing loss, documenting evidence of significant differences between ethnic background and susceptibility to impairment (Lin et al. 2011). Of course, the key risk factor for hearing impairment is environmental and here research absolutely reflects our contributor's insights. Noise exposure is a recognised classification of auditory impairment, titled 'noise-induced hearing loss' (NIHL). As explained by Ding and colleagues (2019), NIHL is caused by prolonged exposure to excessive loudness levels. Acute auditory fatigue, felt from shorter bouts of noise exposure can be recovered from, but there is no absolute quantification of how loud, or for how long, before permanent impairment occurs. Ding and colleagues stress that whilst some treatments exist, nothing can completely reverse the effects of NIHL and therefore a preventative mentality is essential to protect our hearing. Whether it be investing in some decent earplugs, taking an alternate route to work when confronted with jackhammers and drills, or rethinking the appropriateness of standing intimately close to a loudspeaker at a music gig, even sounds that are only mildly excessive in loudness have been proven to have a cumulatively damaging effect. Small exposures add up. Our hearing is most definitely not invulnerable and protecting it is something worth taking seriously.

Affect and identity

Feeling and the emotional aspects of sound

Throughout my many discussions on the meaning of sound, it would be impossible to deny the prominence of its emotional aspects. Without prompts from me, emotion featured throughout the discussions in various different guises. In some instances, emotion was understood by contributors to be one of many, integrated components. In others, it was the central, if not an exclusive aspect of a contributor's description of sound meaning.

I think that I convey most of my ideas and emotions through language and sound.

Ludo: Italy, Egypt, UK

It's definitely about how it makes me feel. Sound has a strong impact on me. I'm very cautious of the sounds around me and I try to avoid certain places where there are certain sounds because I don't want to be in that physiological state of alert and stress. Musical sound is very important in this regard as it relaxes me. I cannot think of a world without music. I feel as though I am constantly trying to balance sound. Music can be positive, but if my son for example turns the music up too loud then it becomes negative. At work, when the sounds are very busy, there are too many people talking, telephones ringing, the printer; then the sound can be very stressful.

Anonymous: Greece

As a teenager, I was into Heavy Metal, which helped me in my emotional development. It was cathartic and helped me become sociable and integrate with a wider social scene.

Stafford: UK

I think that, whereas visuals can be very clean and purely about information, sound is a lot messier. It's more about getting to people's emotions and expressing the emotional content of a story. Especially in film and video games, the visuals prioritise what is happening or how the player is performing, whilst the sound explores the feeling about those things.

Michael: USA

Even objective musical elements can evoke positive and negative emotions. For example, a piece with a tempo of less than 100 bpm, or made of mainly minor chords, or certain instruments such as violin or piano, could elicit some kind of longing from nostalgia or other kind of sadness. There is a subconscious physical element here because we don't have to think about what is happening, the body will automatically respond to certain pieces of music. Non-musical sound can also have significant emotional and physical impact. Certain sounds we could find very irritating to the extent that they make us feel physically uncomfortable.

Andreina: Venezuela

Sound both evokes and expresses emotion. The feelings it instils can potentially influence our behaviour in highly significant ways, drawing us towards something if positive but, equally, driving us to avoidance if negative. Context seems an important factor, with a single listener potentially experiencing very different emotional states from the same sound, depending on the wider situation. At the same time, there is also the suggestion that sound may have more fundamental properties that could potentially be reliable predictors of our emotional response. In most cases, the emotional aspect of sound was described as a rather linear affair; two opposing forces of positive and negative feeling in response to sound. Specifics aside, emotion certainly cannot be understated in terms of its significance. It connects heavily to how sound is experienced and perceived, but also to how we respond behaviourally. Whilst the affective nature of sound is relevant to the subconscious, it is meaningful enough for many to attend to it directly and to consider their actions and even their own routines based on how the sonic environment is making them feel.

Sound and spirituality

As noted earlier, interviewees who presented some of the more expansive responses would typically double as an introduction to the key themes that would characterise later discussions. In these cases, there was certainly a bit of pressure on myself to be responsive to these themes, and to either ask follow-ups that were not on my list of guidance questions or find ways of embedding these themes into subsequent questioning. One notable example of this was the theme of sound and spirituality. This topic certainly made regular appearances throughout my interviewing journey, with the association between sound and religious ceremony and practices being a matter that many of the interviewees wished to offer their thoughts on.

> To me, sound is fundamentally vibration and has the capacity to transform us. Sound can move people, it can change emotions, it can form bonds between people. Sound can transform us in every aspect of life, from politics, to religion, to family and friends. I would say this is more of a spiritual interpretation of sound but, as a vibration, I think that the entity that we can call God, or the universe created us and itself through sound with the Big Bang. I've heard spiritual and religious speakers across various denominations refer to the beginning of creation as 'the first roar', describing the first action being the Word of God. This spiritual concept also translates over to artistic creation. When a musician creates a piece of music, in some sense they are spiritual creators in that they are creating a new world of their own, and that world can potentially hold a lot of power.
>
> Anonymous: Peru

This was a perspective that I found genuinely intriguing but also highly rational. In both ontological and cosmological interpretations, the universe began with sound. The beginning of our existence we describe with onomatopoeia. As Beck (1995) observes, several Hindu scriptures explicitly describe the origin of the cosmos through sound. The sonic nature of the Big Bang is something that scientific literature has considered, notably the work of Cramer (2001), who went so far as to actually synthesise the sound of the Big

Bang. He achieved this by taking the frequency spectrum of the cosmic background radiation captured by the Wilkinson Microwave Anisotropy Probe to render an audio file. Cramer's original 100-second render can be accessed freely online[2] and is certainly worth a listen. Of those who conceptually explored beyond that singularity of creation and ask the question of what came *before*, many religions believe the answer to be an expression of sound. As the book of Genesis, 1:1–3, states: "Then God said, 'Let there be light'; and there was light". The utterance of the word precedes the creation of existence.

Meaning through work

After many rounds of grouping, ungrouping, and regrouping (repeat ad-nauseum…) the many responses to the question of sound meaning, the last theme that emerged from the analysis was that of work.

> It's my life. I've been producing sound now for twenty years. I played the violin and piano from the age of three. I've worked as a DJ and sound engineer, now I would say I am a music composer. In terms of what sound means to me, I don't really know how to express that meaning in words.
>
> Tim: Russian Federation

> For me, sound is a medium for transferring other people's ideas to a broader audience. It is about delivering their expectations, both the creators and the audience.
>
> Gareth: USA, UK

> Sound is my main specialism and is something I strongly connect with my identity. As a subject area and practice, sound is something that I have expertise in, but is also something that I enjoy.
>
> Jon: UK

> It is my job. It's what I do for a living and so it is a big part of my life. I work in a game audio company and sound has become part of my identity.
>
> Can: Turkey, China

Individuals whose responses immediately referenced work often did so with a nod to the feeling of identity. Others described the meaning of sound in ways that heavily related to other themes, particularly affordance and emotion, but largely (if not, exclusively) through the lens of their professional practice. It should be noted that not all sound professionals discussed the meaning of sound within the context of their work, and several contributors elaborated much further on these matters, often exploring instances in which sound featured in unexpected ways within their professions. We return to this in a later chapter.

Chapter 3 summary: the meaning of sound's meaning

Many of us do not actively think about our relationship with sound or how it features in our daily lives. We may also take it for granted. This does not mean that it is not important. If we pause and consider our relationship with sound for a moment, the likelihood is that we will feel that importance very keenly. Whilst arranging the interviews, many contributors explained that their reason for participating was precisely due to their feeling that they never really considered the role of sound in their lives, and that the interviews were an opportunity for them to learn more about themselves as much as it was an opportunity for them to present their insights on sound.

The opening question *"what does sound mean to you?"* certainly proved itself a powerful way to yield a wide range of responses. Whilst this made analysis something of a headache, it helped structure subsequent stages of interviews by inspiring new questions to ask, suggesting certain questions were superfluous and could be removed, and directing me to the better ways of structuring proceedings. Contributors found meaning in what sound enabled them to do, how it made them feel, how it persistently surrounded them, and in how they perceived its value. Some would focus upon the more objective and analytical facets of sound, whilst for others, this opening question served as an ice-breaker between the two of us by enabling the contributor to comfortably explore more personal interpretations of sound meaning. It revealed a great diversity in personal definitions of sound and highlighted that, for many, sound is more a composite of definitions and meanings than something that can be neatly and singularly defined. What I feel to be most significant, however, is a notion that steadily emerged throughout the interviews: that the vast majority of us do not feel that we fully understand our personal relationship with sound, but that with only a small degree of active consideration, we can potentially learn a great deal about what sound means to us, and about who we are.

Notes

1 Results for entering "active passive listening" into Google Scholar: https://scholar.google.com/scholar?hl=en&as_sdt=0%2C5&q=active+passive+listening&btnG= (accessed 18.02.2022).
2 http://faculty.washington.edu/jcramer/BigBang/WMAP_2003/BBSnd100.wav (accessed 24.01.2022).

References

Bauer, C. A. (2018). Tinnitus. *New England Journal of Medicine*, 378(13), 1224–1231.
Beck, G. L. (1995). *Sonic Theology: Hinduism and Sacred Sound*. Hampton: University of South Carolina Press.
Claire, L. S., Stothart, G., McKenna, L., & Rogers, P. J. (2010). Caffeine Abstinence: An Ineffective and Potentially Distressing Tinnitus Therapy. *International Journal of Audiology*, 49(1), 24–29.

Cramer, J. G. (2001). BOOMERanG and the Sound of the Big Bang. *Analog Science Fiction & Fact Magazine*, 1(1), AltVw104.

Ding, T., Yan, A., & Liu, K. (2019). What Is Noise-Induced Hearing Loss?. *British Journal of Hospital Medicine*, 80(9), 525–529.

Fung, C. V., & Gromko, J. E. (2001). Effects of Active Versus Passive Listening on the Quality of Children's Invented Notations and Preferences for Two Pieces from an Unfamiliar Culture. *Psychology of Music*, 29(2), 128–138.

Gaver, W. W. (1991). Technology Affordances. In: Robertson, S., Olson, G & Olson, J. (eds.). *Proceedings of the SIGCHI Conference on Human Factors in Computing Systems* (pp. 79–84). ACM. New York.

Gibson, J. J., & Carmichael, L. (1966). *The Senses Considered as Perceptual Systems* (Vol. 2, No. 1, pp. 44–73). Boston: Houghton Mifflin.

Górska, U., Rupp, A., Boubenec, Y., Celikel, T., & Englitz, B. (2018). Evidence Integration in Natural Acoustic Textures during Active and Passive Listening. *Eneuro*, 5(2), ENEURO.0090-18.2018.

Lin, R. J., Krall, R., Westerberg, B. D., Chadha, N. K., & Chau, J. K. (2012). Systematic Review and Meta-Analysis of the Risk Factors for Sudden Sensorineural Hearing Loss in Adults. *The Laryngoscope*, 122(3), 624–635.

Lin, F. R., Thorpe, R., Gordon-Salant, S., & Ferrucci, L. (2011). Hearing Loss Prevalence and Risk Factors Among Older Adults in the United States. *Journals of Gerontology Series A: Biomedical Sciences and Medical Sciences*, 66(5), 582–590.

Lockwood, A. H., Salvi, R. J., & Burkard, R. F. (2002). Tinnitus. *New England Journal of Medicine*, 347(12), 904–910.

Palmer, A. R., Hall, D. A., Sumner, C., Barrett, D. J. K., Jones, S., Nakamoto, K., & Moore, D. R. (2007). Some Investigations into Non-passive Listening. *Hearing Research*, 229(1–2), 148–157.

Remijn, G. B., & Kojima, H. (2010). Active Versus Passive Listening to Auditory Streaming Stimuli: A Near-Infrared Spectroscopy Study. *Journal of Biomedical Optics*, 15(3), 037006.

Rogers, C., & Farson, R. (2021). *Active Listening*. Eastford: Martino Publishing.

Smith, B. (2000). Objects and Their Environments: From Aristotle to Ecological Ontology. In: Frank, A. Raper, J. & Cheylan, J. (Eds.). *Life and Motion of Socio-Economic Units* (pp. 84–102). Boca Raton: CRC Press.

4 Sound and culture

Gunwharf Quays, Portsmouth. 5th of October 2021. Early evening

I'm standing in the foyer of a cinema. There are only a few murmuring voices in the queue for food and gentle rustlings of popcorn bags and the hiss and gurgle of soft drinks dispensers. The reflective marble flooring and barren foyer create a particular reflective tone that emphasises these otherwise very small sounds. Singular sounds can be clearly identified, such as when a customer accidentally drops a coin on the floor. You could liken the acoustics to a bus depot or train station, but without the accompanying transport noise. The soundscape of a cinema foyer feels unique. This is a pleasant-sounding space. It feels familiar and comforting, but with shades of nostalgia and even excitement.

In the previous chapter, we looked across the numerous and varied responses that our contributors provided concerning what sound meant to them. As the narrative of each interview progressed, a great number of contributors felt comfortable enough to take the reins. This meant that, in many cases, asking a prescribed second question was not appropriate as the interviewee had landed upon a particular topic about sound that they instinctively wished to explore. Fortunately, the matter of how they perceived their culture to be an influence on their relationship with sound proved relatively straightforward to include within most interviews as, irrespective of the specific points of the discussion to that point, it was possible to reframe those specifics around the question of cultural influence. Unfortunately, the limitation with this approach to the question was that the response would largely be restricted to that preceding topic. If the previous discussion centred upon work, or home-life, or language, and so on, the matter of cultural influence would typically also centre upon the same theme.

The aim of these questions on sound and culture was to examine the potential for there being a 'cultural component' within human auditory perception

DOI: 10.4324/9781003178705-5

that meaningfully influences our relationship with sound; how we interpret, interact with, and draw meaning from sound, and how this affects us in our everyday lives.

Themes of culture and sound in research

Scientific literature, by and large, agrees that human auditory perception is meaningfully influenced by both universal principles and cultural factors (Wong et al. 2012). Certain acoustic effects, or the hearing of particular sounds in specific contexts will be interpreted by us humans irrespective of who we are, where we are, and where we are from. However, research is discovering more and more ways in which culture is heavily influencing how we interpret, value, respond to, and utilise sound.

How should we be defining 'culture'?

One of the first things that emerged in the early discussions with contributors, was that the term 'culture' brought with it some challenges concerning definition. *"How do you define 'culture'?"* was asked a couple of times and is a fair question. Although academic discussion on the subject of culture goes back much further, a 1957 review article by Goldstein concisely brings together a range of perspectives on the matter. Goldstein cites Omar Moore's (1952) writings upon anthropological methodology, arguably relevant to this book and its ethnographic ambitions, positing that researchers cannot rely upon a single catch-all definition and therefore must always "state explicitly and unambiguously what they mean by culture" (Goldstein 1957: p.1075). Goldstein continues, introducing us to various approaches previous researchers have taken in search of a meaningful definition, all of which appear to have come up short; riddled with gaps and inconsistencies. Fortunately, the ethnographic approach rides in to rescue us with the reminder that our process is fundamentally 'interviewee as expert', and that the control over how culture is defined must be established individually by the contributor. That said, it would be remiss of us not to take a brief look at studies into auditory perception that specifically address cultural factors, to help give ourselves some initial guidance, and to avoid placing all the responsibility for defining culture upon the contributor.

Earlier research

Studies comparing cultural influence on auditory perception broadly represent culture as synonymous with ethnicity. Experiments go back at least as far as 1965 with a study by Birch and colleagues. This study now makes for rather uncomfortable reading as it compares listening accuracy across racial lines within the historical context of the Civil Rights Movement. A few years later, a 1971 study by Breger revealed differences in 'pleasantness' ratings of 12 sounds between five cultural groups, identified as: "Afro, Anglo (English-speaking), European, Israeli, and Middle Eastern" (p.315). In 1979, an experiment by Shigehisa compared cultural influence across religious delineation, observing that, when exposed to lights and tones of varying intensity (brightness and loudness, respectively), Buddhist participants rated intensities of both light and tone significantly greater than non-Buddhists. This led Shigehisa to surmise that self-attainment could actually make an individual more sensitive to loud sounds, implying that cultural effects are in fact a complex web of psychological factors that influence auditory perception.

More contemporary perspectives

In the last two decades, research has pointed to various ways in which 'cultural difference' impacts on the human relationship with sound. For example, functional magnetic resonance imaging scans have revealed that individuals with mono-cultural backgrounds actually use different parts of the brain to differentiate types of music when compared to bi-cultural subjects (Wong et al. 2009). In 2012, Wong, Chan, and Margulis suggested a combinatory effect of bottom-up and top-down processes that collectively form our cultural relationship with sound. Here, bottom-up describes the literal shaping of our auditory-neural pathways based upon the sounds we are repeatedly exposed to, whilst top-down refers to our cultural upbringing and how our expectations, interpretations, and preferences in the world modulate how we perceive sound. Research concerning top-down influences on perception predominantly focus on vision, though Wong and colleagues assert that factors influencing visual perception could also apply to sound. One example of this is the theory that collectivistic cultures, featured more prominently in Eastern societies, balance their focus on an image between the background and foreground, whilst a more individualistic (typically Western) culture would focus heavily upon the foreground. This theory has already been evidenced in the visual modality (Chua et al. 2005), but it is not a great leap to imagine that it would also apply to sound.

Though not exactly a comparison of background vs foreground, a couple of years later, Cao and Gross (2015) compared loudness perception of tones under three conditions: (1) triggered by the participant; (2) by the researcher; and (3) by a computer; and between Chinese and British participants. Their results revealed that whilst British participants gave higher loudness ratings to sounds triggered by others compared to those that were self-triggered, Chinese participants showed the reverse, perceiving tones triggered by others as quieter. Cao and Gross considered their results against what is known as the 'internal forward model', the theory that when a person performs an action, they also generate an 'efference copy' of that action. The efference copy is basically an imagined representation of the action. I open a door and, simultaneously, I generate the idea of myself opening that same door. The efference copy in turn generates a 'reafferent response signal'. I imagine the result of my opening the door as I am opening it, continuously updating a perceptual model of what I expect to happen. The point of all this is that the experience of that reafferent signal at the same moment as the action itself prompts my brain to attenuate the intensity of the actual sensory experience, helping me to tell apart self-generated stimuli (or feedback) from external stimuli that I am not controlling, ultimately providing me with a sense of agency. Bringing all this back to the results of Cao and Gross' study, the fact that Chinese participants did not perceive a less intense sensory experience implies that the internal forward model is less applicable and that this is an effect connecting to Eastern collectivist culture.

In a systematic review of 19 separate case studies at urban sites across Europe and China, Yu and Kang (2010) extensively probed the question of what factors influence sound preference in the context of urban spaces. Their review concisely groups factors into four groups: environmental, behavioural, psychological, and social/demographical. Environmental factors referred primarily to time of day, climate, and local weather when visiting the space. Behavioural denoted participants' reasoning for visiting. Psychological described overall liking/disliking of the space, whilst social, and demographical factors grouped together age, gender, occupation, education level, and residential status (whether a participant identified themselves as 'locals'). The sounds themselves were grouped into three classes: natural (bird song, flowing water, insect activity), human (speaking, footsteps, children shouting), and mechanical (vehicle/traffic and construction). This factorial design created a lot of relationships to examine and, consequently, a lot of data which we can't unpick in detail here. However, what is worth noting is that there were a number of statistically significant correlations drawn between sound preference and several of these influential factors. As age increased, so too did preference for bird and insect sound, and greater dislike for the sounds of shouting children. Higher levels of education on the other hand showed a more negative response in general towards mechanical sounds, perceiving them as noise more readily. According to the data, women from Shanghai preferred bird sounds more than men, whilst in Sheffield, the reverse of this gender difference was observed. Although culture is not explicitly defined within their review, Yu and Kang use the term broadly to reflect different geographical locations and where a participant identified as 'being from'.

Whilst rarely presenting an explicit definition of culture, research studies into auditory perception over the decades have rarely deviated from implying by the designs of their studies, that culture is synonymous with a participant's country of origin. It is also notable that, when comparing cultural effects in these terms, the studies by and large compare Eastern and Western countries. In evaluating native English and Japanese speakers, Iversen and colleagues (2008) demonstrated a significant difference between the groups in terms of how they perceptually clustered rhythmic sequences of tones. In 2012, Wong and colleagues explored cultural effects on musical pitch perception, in this instance comparing participants from Hong Kong and Canada. There are exceptions to this rule of course (e.g. Walker 1987; Fonseca 2014), but these appear to be the minority.

Initial thoughts from the interviews

In practice, preliminary interviews revealed that many contributors were a little uncomfortable with defining culture themselves. When asked the question of cultural influence upon sound, several immediate responses were to ask me pointedly to define culture, so that they could frame their response within the restraints of my definition. Returning to the earlier observation that much of the research concerning auditory perception tends to associate culture with country of origin, this did seem like a good example to state first, but it did not feel appropriate to leave things there. Many of the studies above compared the effects of more than just country of origin. As such, it actually made a lot of sense to acknowledge some of these broader social, individual, and demographic elements such as age, gender, personal history, generation, phase of life, education, class, and so on. I must admit to not being particularly consistent with which set of features I would use as examples between interviews, but I was sure to always end my clarification of the main question by making it clear that "*it can be a lot of things, but you are free to define 'culture' in whatever way you feel is most appropriate*".

Short and sweet responses

What became apparent very quickly from the interviews was the sheer scope of how 'culture' could be defined by the contributor. Some considered influences along more national lines, others thought even broader and drew comparisons between Eastern and Western societies, whilst some went the other way and considered their local or personal history as the most relevant factor of cultural influence.

> Culture absolutely influences sound. I think of the emotional memory of my life. All the sounds I hear creates a soundscape which resonates with my style of living, how I feel in my life, my fears, my happiness, everything.
>
> Heikki: Finland

> Culturally, we are all affected differently by sounds. You might draw different associations to different sounds depending on where you grew up or what traditions you followed. Music is a very good example of this, where different cultures can have different scales, that to other cultures might sound 'foreign'. So, for some, a scale could sound 'safe' and 'homey' while for others it could sound like it's 'unknown' or 'different'.
>
> Rasmus: Denmark, Sweden

I don't think that cultural factors play a significant part in our relationship with sound. I think that sound is sound.

Jahangir: Bangladesh, UK

I haven't thought about cultural influence on my understanding of sound
before. Growing up in a large city in India, when I moved away, I found
everything much quieter. In that respect, the absence of sound has played
a big part in me understanding the importance of sound.

Anonymous: New Zealand, India

As a Spaniard, I would say that I'm very used to noises, everywhere. How
I think about sound goes back to my professional background as a sound
designer for theatre and now going into research.

Angélica: Spain

In terms of broad observations, the majority of contributors clearly expressed
a belief that cultural influence does indeed affect our relationship with sound.
This was, however, not a unanimous position, and it should be noted that
several of the contributors did not feel that culture made a difference. Of
those who did assert to a connection between culture and sound, we can see
references to different countries, personal backgrounds, work, and persistent
acoustic environments as specific elements of focus that instinctively came to
mind for our contributors when posed the question, suggesting that 'cultural
effects' best describes a wide range of factors that can influence auditory perception in different ways, and to varying extents, on a person–by–person basis.

Detailed, and a little more intimate

On the topic of sound and culture, many contributors preferred to put forward more of an overview, briefly considering several potential ways in
which culture may influence their relationship with sound. Though more of
a minority, a good number of interviewees intuitively, and without further
prompt or pressure from me, focussed upon a specific theme and elaborated
on it. These responses were too long to be considered short yet, somehow,
also too short to be considered long. Either way, discussions often began to
feel notably more intimate at this point in the interview, as if we had crossed
a distinct line between detached formality and personal insight.

Whilst I grew up in the UK, my family used Sri Lankan film and television to teach us the Tamil language, and about the culture. Film was a big
part of my life and music was a big part of the films. A lot of these songs
are about community, family-based messages and that is a core value of
Sri Lankan culture. If I want to cry or feel a particular emotion, I will
listen to certain types of songs. Tamil songs in particular give me a real

sense of home. Living in the UK, these songs give me a sense of more personal identity and heritage.

Laksha: Sri Lanka, UK

I think that the meaning of sound is very different for different people. Japanese is a culture that is close to nature. This emerges from a mix of religion, farming culture, and a history of significant natural disasters such as tsunamis, typhoons, and earthquakes. In Japan you have two major religions, Shinto and Buddhism. Shinto is based on a respect for nature. So, sounds and nature and life are all connected. Sound is always there. It tells us the seasonal changes or even the time of day. Historically, Japanese industry has always involved a lot of grain and vegetable production. This has meant many people have passed down from generation to generation the understanding of the climate and the seasons. It's quite common to teach young children these listening skills in school.

Shoko: Japan

As can be observed in the above responses, family appears to be a key element that, for many, can act as a central core, drawing in other cultural elements to determine an individual's composite 'cultural processor'. We can see how family features as a means of handing down awareness of particular sound affordances (in the above case, the affordance of sound to inform upon the climate and the season), but also how to develop practical listening skills that enable the listener to effectively realise that affordance. Family can influence exposure to particular musical styles and genres, connecting us throughout childhood to particular emotions, messages, and values.

I think that there's so much noise in Spain that it's difficult to find your own sense of sound. The quieter space in the UK gave me the space to explore the sounds a lot more. I found a choir in the UK, and this made me think about how many smaller sounds could amount to something much bigger. I appreciated more the idea of sound being spiritual and I was encouraged to actively listen to natural sounds such as the song of birds or the wind in the trees. First it was a process of finding my sound. Once I had found it, I understood the power of these sounds to me.

Lucia: Spain

Iranian expression is softly spoken. I think culture can influence the sound of speech. Arabic, for example, is naturally vigorous and energetic. I think cultural background can influence how we interpret the sound aesthetics of different languages. So, one language is not objectively more pleasant than another, but your background will influence which languages you prefer the sound of.

Mahmood: Iran

Music is a big part of my life. I have a musical family and I enjoy music as a consumer. I do enjoy silence, and conversely can get annoyed by too much sound information. I have three rather small children so I could definitely benefit from a little more peace and quiet! I do find that I enjoy breaks from sound and music; maybe because I'm very used to attuning to sound and am usually processing it very actively, which can be tiring, and I don't have the attentional bandwidth to manage that constantly without a break.

Brecht: Belgium

As with the earlier responses, family emerges once again as a theme, but here it is exceeded by the notion of sound-energy. This is expressed predominantly in relative terms of loudness (i.e. louder, quieter, noisier, softer, etc.) but also velocity (vigour, intensity). With regards to cultural influence, the emerging principle appears to be that the historical energy levels of your sound environment will benchmark your response to sound in the here and now. We acclimatise to the energy of sound over time, be that our language, the music we listen to, the ambient soundscapes we experience in our homes, workplaces, or out in the world. This acclimatisation creates expectation and a personal impression of 'standard sound' against which we evaluate (albeit, largely subconsciously) current sound.

When I was younger, I remember watching a behind-the-scenes documentary on the series *Friends*, discussing the use of foley on the show. I found it very intriguing, the idea of all the incidental actions like footsteps, clothing movement, and picking up objects, being recreated in sound in post-production rather than being captured during filming. Once I saw that, I found myself watching other television and films and thinking about the process behind the sounds on screen.

Meni: Mexico

The previous response is one of my personal favourites. The notion of a seemingly innocuous occurrence acting as a pivotal moment in a person's timeline that would have far-reaching implications for their future. Of course, although such a narrative is compelling, almost poetic, there is a bit of a risk that such an observation is suffering from fundamental attribution error. Heider first put this concept into words in 1958 (reprinted in 2013), suggesting that humans have a pesky tendency to falsely observe cause and effect where there is none. Attribution error may be situational (when assigning the cause of behaviour to an external situation or event) or dispositional (when the cause is believed to be the result of some internal characteristic or personality trait). A more recent review by Maruna and Mann (2006) explains that this cognitive distortion can have various additional dimensions. We may prescribe intent where there is none. We may mistakenly identify specific causes when the originating factors of a behaviour are far more global or distributed.

We may also incorrectly presume the cause of a behaviour in one context will be consistent (or 'stable') over time or in a different context. Fundamental attribution error is an unavoidable element in our consideration of all the responses to this question of sound and culture, but it is important to emphasise that what we are asking specifically concern's interviewee's *belief*, not their factual knowledge. Indeed, the belief that something is an important influence on our decisions may in itself imbue that something with considerable influential power. No one is suggesting that every person who sits down to watch a 'making-of documentary' will be inspired to work in that industry, or that, in the above case, there were not a whole host of additional variables that unwittingly conspired to comprise the emergent decision to pursue a career in sound. But this does not mean that such an event was not significant. Again, the intention of this book is not to answer questions, but to generate them. Not to prove causality or even correlation, but to raise it as potential.

Emerging topics for further discussion

The sound of coffee

Believe it or not, in the writing of this book, and at the grand old age of 37, I actually 'discovered' coffee. Now my very own everyday sound environment is punctuated by the clattering crunch of my (apparently) amateur ceramic burrs upon a handful of coffee beans and the bubble, hiss, and splutters of the Bialetti Moka pot. This is on days where I don't rely on my automatic espresso machine, which can be best described sonically by its shuddering groan, as if it were chastising me for expecting coffee at the press of a button. In this instance, auditory perception may be saying a lot more about me than about the acoustic properties of the sound – but is that not the point?

This seemingly arbitrary aside does in fact provide what some might consider a rather unusual but, nevertheless, effective subject for exemplifying the relationship between culture and sound. According to Spence and Caravalho (2020), an estimated 4 billion cups of coffee are consumed every year. This consumption is not distributed evenly across the world. For example, a review by Euromonitor International,[1] states that in 2013, whilst China and Uzbekistan effectively brewed no coffee whatsoever, the average consumption in France, the USA, and New Zealand clocked in at roughly 60 litres per person across 12 months. The Netherlands topped the charts with almost 240 litres. In a recent review of the scientific literature, Ghahraman and colleagues (2021) revealed that caffeine consumption can have a profound effect on our hearing, identifying a correlation between drinking coffee and lower incidences of hearing loss and tinnitus. They also presented studies that evidenced caffeine could reduce the latency and enhance the amplitude of sound signals as they progress through the auditory system, suggesting that caffeine consumption can impact the speed with which we respond to sound cues, but also our sensitivity to soundwaves in general. If we consider these effects in the context of varied distribution of global coffee consumption, we now have the recipe to hypothesise that some countries may experience sound differently at a fundamentally physical level, due to variations in how much coffee they consume as a nation. Furthermore, the potential picture of sound and culture, within the contextual frame of coffee, may be even more complex. According to Knöferle (2012), if I were to filter out the high frequencies emanating from my aforementioned coffee machine, presumably with some suitable earplugs, the subsequently produced coffee would taste sweeter and less bitter. *However*, based on the findings of a 2018 study by Ong and colleagues, the bitter taste counterintuitively encourages more coffee consumption due to a sense of positive reinforcement; bitterness implies caffeine, caffeine implies alertness, and alertness is the primary driver for drinking the coffee in the first place.

Now I'm confused as to whether I should use the ear plugs or not whilst preparing breakfast tomorrow. What is more apparent is what can best be described as an ecological structure of sound and culture, here in the context of coffee. An acoustic ecology of coffee, in which sound affects our perception of coffee, whilst equally, coffee affects our perception of sound. This notion of an ecological structure is important because, as we move onto the more extended interviewee responses to the question of culture and sound, the discussions start to reveal numerous potential connections, suggesting that none of the observed cultural effects exist in isolation, but rather as a far more complex, interconnecting web of effects from which a singular experience of sound emerges.

Synaesthesia

Of these detailed responses to the matter of sound and culture, several raised a particularly fascinating phenomenon that can be an extremely significant part of the cultural processor: synaesthesia. In a 2012 review, Simner traces synaesthesia back more than 200 years, with the concept first put to page in 1812. Though the traditional definition of the term has been rigid for many years, Simner argues that this situation is changing and that a more contemporary understanding that accommodates more recent developments in neurobiology is required. Prior to this, synaesthesia was generally understood to denote a 'merging of the senses' which, over time, constructs long-term perceptual connections between two or more sensory modalities. The term itself is a composite of *syn-*, meaning joining, and *-aesthesia*, meaning sensation. The merging of the senses can, in the most basic terms, be described as responding to a physical stimulus of a particular sensory modality, vision for example, with a perceptual experience that is both visual and (again, for example) auditory. In this understanding, it is fair to view the modality that has a direct physical correlate as 'actual' and that which does not as 'imagined'. Putting aside the potentially patronising connotations of synaesthesia being an imaginary experience, many synaesthetic individuals do not feel that this depiction is accurate. Simner supports this opposition, arguing against this notion of merging senses, and pointing to more recent evidence that shows many synaesthetic experiences are not triggered by a physical stimulus, but by the high-order percept, it corresponds to. Simner provides as an example, citing several studies (e.g. Smilek et al. 2001), the experience of the letter *a* as the colour red, noting that the visual depiction of the letter upon the page can be heavily altered and still trigger the experience of the colour red, provided the letter evokes the idea of the letter *a*. Here, it is not the physical image that triggers the synaesthetic response, but the perceptual construct (the idea, or feeling, or notion) of *a* that incorporates colour into its collective meaning. Essentially, a perception generates an accompanying perception. This may be an experience of two perceptions of different sensory modalities, but is more likely, according to the research, to be of different facets within a single modality. For example, rather than say, seeing colour in a sound, a more commonly reported example of synaesthesia would be colour in a shape.

> When I think of works by Eric Satie for example, I feel that is very plastic music in a way, very sculpted and vivid. It is music that you can almost touch. That's something that has been with me since very early childhood. I think cultural influence is itself synaesthetic. It's not just about the sounds, it's about the smells, the tastes, the pace, the colours of the environment. All mixed together. That makes my culture.
>
> For me, sound travels visually. To a large extent I would say my experience of sound is synaesthetic with vision, but also other sensory modalities. The experience can get messy; fractals calling other fractals in all directions. A complex fractal feedback loop. I would say that I experience colour from sound more strongly than some other synaesthetic experiences, but it can take time to immerse yourself in a synaesthetic experience. I also find that the reverse, experiencing sound from images, works in this way. I find that the synaesthetic experience also occurs in everyday listening. When I see certain actions or events but don't hear a sound, I will always feel some kind of sound for it.
>
> Nicolas: Iceland, France

Nicolas' description of sound and culture through a synaesthetic lens as messy, complex, fractal, and vivid certainly reflects the scientific literature, both in terms of the synaesthetic experience being too visceral to compare with an imagined experience, but also in synaesthesia being a complicated, multi-layered, and fluid phenomenon.

> Sound can have an image perceptually attached to it. It can have a taste, or a bodily feeling. When you're born as a synaesthetic person, it is not different to you, it is just the way in which you see the world. Then, at some stage, there is the realisation that you're experiencing things in a way that not everyone else does. I remember noticing a poster in the

university canteen from the psychology department, calling for participants for a study into synaesthesia. It was liberating to realise that there
was nothing wrong with being synaesthetic.

   In terms of my experience of sound in relation to synaesthesia, I would
say it is possible to experience both sound as non-sound, and the reverse
of non-sound as sound, but it is more likely that I will experience the former. For example, I can also tell you that, in this interview, the sound of
your voice is a dark red, Bordeaux Red to be precise. There's no guarantee that it will be the same colour tomorrow. I have in many cases found
synaesthesia to helpful. One example is that I find it easier to remember
equations because of their synaesthetic sound and colour. The experience
of synaesthesia is most similar to a memory. More visceral and immediate
than an imagined percept but not equal to a sensed experience.

Anonymous: Italy

In a 2019 positional paper, Ward explains that synaesthesia can be developmental (emerging throughout childhood as the brain grows and matures),
acquired (as a compensatory adaptation of the brain in response to sensory
loss, or due to psychoactive substance use), or intentionally trained. As Ward
asserts, synaesthesia has a distinct phenomenology (the nature of its felt experience) and unique set of causal mechanisms which means synaesthetic
individuals should not be positioned along a neurotypical continuum. Essentially, it would be inappropriate to suggest that we are all synaesthetic
to some extent with some simply being 'more so' than others. However,
whilst not exactly challenging Ward's argument, one particularly intriguing
theory doesn't quite fit is the 'neonatal synaesthesia hypothesis'. This theory
resonates heavily with our discussion on the ecology of sound and culture. It
suggests that we are all born synaesthetic, effectively experiencing the world
as a singular sensory modality, but that we 'compartmentalise' our impressions of the world into modes as our brains develop throughout childhood.
Attempts to validate this theory have yielded mixed results. A review by
Kadosh and colleagues (2009) suggests there is more support against the theory than for it, but this is not enough for them to take a conclusive position
either way.

   Interestingly, an earlier paper by Simon Baron-Cohen (1996) explains the
difference between neonatal synaesthesia (NS for shorthand) and the related
theory of cross-modal transfer (CMT). Baron-Cohen observes that, unlike NS, CMT has amassed a substantial quantity of empirical evidence, is
more widely accepted throughout the discipline, and is decidedly less radi-
cal. Whereas NS suggests that all infants experience the world as a singular
stream of sensory input, CMT instead posits that they recognise objects in the
world across multiple modalities as a result of being able to form higher-order
cognitive representations. At this point, I am compelled to point out that I am
in no way an expert in synaesthesia, or even developmental psychology, but
I could not help but draw a parallel between Baron-Cohen's description of

CMT in infants and Simner's (2012) perspective on synaesthesia as a perception that generates an accompanying perception. On this basis, my takeaway from all this is that we may indeed all be born synaesthetic, but exclusively in terms of cognitive perception and *not* physical sensation. A photon and a soundwave are not the same, nor is a particulate that evokes a sense of smell equal to the material that evokes a sense of touch. However, our experience of the world is cross-modal. Most of us perceptually reinforce distinction between the senses as we grow, but this is a developmental process that could potentially be overdeveloped, encouraging us to focus on the individual elements within a system, rather than observing the connections between them and the system itself more holistically. Furthermore, in many situations, this may not be optimal and the deliberate development of cross-modality could yield benefit to how we experience and interact with the world.

Music and culture

The focus of this book was always intended to be on non-musical, non-speech sound and I was keen to avoid any discussions being hijacked by music. However, to be fair to our contributors, music and culture are hugely intertwined. To expect music to be absent from a discord where you are exploring the relationships between sound and culture would be a little ridiculous. This very book dedicates more than a couple of pages to a consideration of ethnomusicology and, as Herndon and McLeod (1981) attest to in the very title of their study on the subject, we could think of music *as* culture.

> I think that my experience with music when I was young has been very significant. My family are descendants of Spanish ancestors but have strong African roots. Culturally, Peru is a deeply mixed country, bringing together the indigenous Peruvians with Spanish and African settlers. I studied music in Spain and lived there for many years. We have several significant musical styles that, reflecting our culture, are very fusion based. We have Afro-Peruvian as a key musical style. Sound and music have always been a way for me to feel as though I am connected to something. In such a mixed culture, I didn't feel that I had a clear sense of identity but exploring this music helped me to form my own identity. I felt that this music belonged to me. It brought me closer to my own culture. I also grew up listening to Reggae. My family enjoyed this style of music, and I grew very fond of it as a child and began feeling increasingly connected to the Rastafari culture through the Reggae music. I remember discovering a Moroccan culture called Gnawa. It is a form of traditional music and rhythms that capture the pre-slavery culture and history, mixing this with Muslim influences. When I first heard this music and learned about its history, I was very moved. There were many connections to my existing cultural background, and it really resonated in me. I found that Reggae and Afro-Peruvian music were

strongly influencing my work. These styles felt very primal, connected
to the body and to the heartbeat.

Anonymous: Peru

Here, we can see that music is deeply connected to identity, but also to family
and work. From the insights presented above, it seems appropriate to consider
music itself as a connector; a bridging device that can connect us to multiple
aspects of culture and community. These connections appear stronger when
multiplied, such as above, where the feeling of identity through music appears
reinforced by associations between music and family, work, and national heritage, but also connections to both the human body and to key moments
in personal history. Such connections are reinforced by scientific literature.
Back in 1997, Ruud conducted a thought-provoking study in which participants were asked to produce a mixtape of songs they felt had significantly
impacted upon their lives, plus an accompanying personal reflection on this
"musical biography" (p.3). Ruud realised that it was possible to build an explicit theory of music and identity comprising four key assertions: (1) music
making offers the individual an opportunity to be 'seen', by a parent, a peer,
an audience, or a community; (2) music affords a social space than can enable
an individual to become embedded within a social group, to attain a feeling
of 'membership' to class, subculture, gender, sexuality, or ethnicity; (3) music
intrinsically evokes and reinforces a sense of time and place, forming attach-
ments to both geographical locations but also points in history, marking both
key events and transitions between life-phases; (4) music evokes physiological
sensations, encouraging proprioceptive feelings of physical self-awareness and
a connection between the self and the body; and (5) music can potentially
extend beyond spatial, temporal and cultural boundaries to explore transcendental and transpersonal space within which an individual may be able to
connect with something "indefinite and indescribable" (Ruud 1997: p.11).
Much of Ruud's thinking resonates with contributors' responses.

If I think of sound, I think of music. I think music is the most profound
way to explain the inexplicable. It can make things both objective and
subjective. It can convey truth. In speech, one person may have a thought
that they wish to convey. That thought is translated into language and
expressed to another person, who tries to reconstruct the thought. Of
course, the end result of each person's experience has to be different,
even if the two people are very close to one another. I think that music
is the medium that circumvents this problem altogether. Music doesn't
even attempt to translate or communicate an objective meaning. It puts
your own feelings, in some inexpressible and untranslated way, into a
pool that someone else may swim in; leaving it open to any interpretation
yet, paradoxically, giving us the opportunity to experience its original
meaning.

Anonymous: Italy

For me, music is very much a cultural thing, and something that impacts my relationship with sound. I identify with certain types of sound and music. My early experiences with computers have also shaped how I think about sound. I grew up working with systems like the Atari ST, and I feel very connected to the idea of using computers to be creative and play with sound. I do think that there is a dimension which links aspects of identity and interpretation of sound and music. For example, I like the idea that shared enjoyment of genres like drum & bass, which have specific associations with dub reggae and Black-British culture, may help cultivate a positive sense of multicultural society in the UK. I would also argue that representing a spectrum of genres is important in university music curriculums because you are potentially giving greater access to students from a wider range of backgrounds to investigate specific areas of music that they feel represent them and their tastes.

Jon: UK

I was a middle-class child of the '70s and '80s, born and raised in Wales. I would say that sound would only really feature creatively in schooling through music class, or in a very dry, factual delivery in a physics lesson. My home was always filled with music growing up. My dad always used to play the most amazing concept albums such as *War of the Worlds* and *Journey to the Centre of the Earth*. My siblings were a bit older, so I was influenced by their interests in '80s popular music and electronica. Games technology was also an important factor early on. I used to use a noise-tracker programme on the Atari ST to sequence drums and, from there, learned the fundamentals of computer-music composition but also the processes of producing my own sounds, making samples then altering them. This started me on the path towards my game audio career.

Stafford: UK

Music is clearly a powerful aspect of sound's affordances of emotion, mood, relaxation, and connection, but arguably extends further across further elements of sound's function and value. As can be seen above, music can also emphasise technological aspects of culture. An individual may be introduced to a new technology by way of its application including the creation or playback of music. The music we are exposed to throughout our lives has been persistently dependent upon the technology of the place and time, encompassing everything from the exclusivity of the harpsichord to Western Europe in the 18th century to the, even today, deeply uneven global availability of the smartphone (Wood 2019).

I think there is a snowballing effect because social interaction and culture are deeply connected, and music invites social interaction through shared passion in something. Through music and people, we would share music, talk about music, explore, and search for new music together. I have found that there is a cultural difference when it comes to perception of live music, particularly street performances. In the UK, certain music is part of cultural identity. So, if I were to sit on a bench and play *Blackbird* by The Beatles, passers-by would often make it clear that they were impressed and perceived me to be very cultured. If I were to play the equivalent in Italy, people would be far less responsive. It's not that music is not important to Italy, but rather that it is part of the common fabric of everyday life and so one person, performing on the street wouldn't typically stand out. Music and musicianship are very common in Italy but more so as a hobby than as a profession. Here in the UK, I have the feeling that music is truly part of your cultural identity, of being British.

Ludo: Italy, Egypt, UK

In addition to reinforcing several earlier points regarding music and culture, the above quote also presents a particularly interesting observation: that different cultures across the world may ascribe decidedly different levels of value to music, specifically live music performances in towns and cities. Aligning with the complex natures of the various elements discussed throughout this chapter, live popular music in cities has a profound cultural component and distinct ecological structure. Research has observed relationships between the perceived value of live music and expectations for economic impact in terms of tourism, consumer spending and job creation (Carter & Muller 2015). Additional factors include political priorities, subsidies for music venues, and even laws concerning alcohol consumption (Homan 2019). Of course, these factors vary dramatically from country to country, supporting the idea that the perceived value of live music cannot be uniform across the world. Consequently, the soundscape that immerses us within our own towns and cities, at least in terms of live street music performances and our attitudes towards them, is likely to be unique to that place.

Local, historical, and seasonal effects

The effect of place on our relationship with sound is something that I strongly expected to emerge as a prominent feature across the interviews. How geographical location around the world could affect the soundscape of our personal environment, but also our interpretation of sound, was a question that I felt personally invested in exploring. Fortunately for me, several contributors began to highlight relationships between sound and culture within this theme of place.

A sound is just a vibration in the air that one can perceive and interpret. But our relationship with sound, the way we perceive it, and the meaning we get from it can vary from culture to culture. Depending on your background, a sound can have a particular meaning, or possibly no meaning at all. There are historical effects on different cultures in terms of sound that make me think of ways in which other places find meaning in sound that I don't. For example, countries such as Japan, that were bombed during the Second World War, still have a deep emotional connection to the sound of their air raid siren, and I am conscious that this is a sound-meaning that we don't have in my culture.

When I was younger, I used to become a little stressed towards the end of the year, as Christmas was coming. People would call this the 'end of year stress'. I didn't really understand where this stress came from, but I now believe it's connected to expectation and wishing for everything to be perfect. Then this stress gets effectively transmitted to other people in a family. I remember feeling it growing up in my own family. What's interesting was how this feeling would be triggered by sound, particularly the playing of Christmas music in the markets.

Anonymous: Senegal/Togo

Two observations in the above response piqued my curiosity. The first was the reference to air raid sirens. For me, this raised a question concerning acoustic variation in alarm sounds between countries. A 2017 review by Carmel and colleagues reveals that alarms can broadly be characterised by either pulses, frequency modulations, or alternating tones. Otherwise, variations beyond these generalisable acoustic properties are commonplace, and the researchers observe that although an international standardisation framework for auditory danger signals does exist, it is not widely used, with most countries possessing their own model for alert sounds. This has been emphasised within contemporary 'meme' culture, in which several online content creators have produced 'ratings videos' in which they audition numerous alarms from around the world. Whether you should or should not find the presentation style of these videos entertaining (or even palatable) is not for me to say but hearing the substantial range of acoustic differences between the various sirens is something I undeniably found fascinating. I would recommend watching at least one of these videos.[2] How different cultural factors might impact a person's response to sound across a range of instances (from pedestrian crossings to vehicle-reversing warnings and beyond) is explored further in the next chapter, where we review sound and place in more detail.

The second piquing of my curiosity came from the point concerning Christmas. The notion of seasonal anxiety being triggered by 'the sounds of Christmas' was an observation that I found particularly fascinating and was encouraged to trawl the literature for references. A few studies do sit around the periphery. Morrison and colleagues (2011), for instance, revealed that shops utilising Christmas scents to persuade shoppers to linger and purchase should ensure they also play Christmas music. In tandem, the scent and sound had a significant positive impact on customers (at least from the perspective of the retailer), but a mismatch would be more likely to drive customers away than to entice them in. A somewhat unnerving 2001 study by Merckelbach and van de Ven sat participants down with an extended recording of white noise and a single button to press. Listeners were instructed to press said button whenever they heard *White Christmas* within the noise. At no point in the white noise recording did the song actually play. In all, 14 of the 44 participants pressed the button at least once. Unfortunately, the researchers did not repeat the process with alternative, *not-really-there* pieces of music to establish whether Christmas music actually stimulates hallucinatory responses more than other genres, and yet I still find this study disconcerting the more I think about it.

Although not precisely the focus of the work, Ronald Lankford's (2013) cultural history of American Christmas songs does reveal that numerous modern holiday favourites are lyrically laden with expressions of anxiety regarding the expense of the season, loneliness for single adults, and dealing with 'stick-in-the-mud types' who don't appreciate the 'true (commercial) spirit of the season'. Based on the evidence available, there are certainly some strange goings on in the world of Christmas sound, but more discoveries remain hidden in the white noise.

> I would expect that, in terms of cultural aspects, there will be things that I hear in my everyday life that you don't in your own. The sounds that I hear on a daily basis give me a sense of where I am, they even tell me about whether I still want to be here or not. I'm a diver, and one particular way of de-stressing for me is to go underwater. Listening to myself blow bubbles and listening to the diverse range of sounds, from the underwater silence to the sound of the parrotfish eating away at the coral. When I go into the ocean and I just listen, I can hear all the marine life just eating away.
>
> Anonymous: Malaysia

> I think, from a cultural perspective, that levels of sound are different, and I think that has an impact on my relationship with sound. Norway is a quiet country. It's a very peaceful place, and to me, the sound reflects the social and political peacefulness of the country. I'm totally generalising here, but Norwegians like to be out in nature. They like to be away from the city, where there is no noise, no traffic. All you can hear is birdsong, the trees in the wind. By comparison, the UK feels like more of a noisy country. I feel a lot more aware of sound in the UK because I like things to be quiet and I suspect that's a result of being more acclimatised to the quiet, simply because where I grew up. Sometimes I wonder if it's not really the sound itself, but rather the idea of a busier, high-pitched "city-ness" that I feel affects my interpretation of the sound. I think of pollution almost as being noisy, as if the idea of it in the world adds to the background sound. I'm not literally hearing it, but it is making everything feel noisier. It's difficult to describe but I think that sound is more volatile in the UK. I think there are more layers to the sound. These layers are always overlapping, and it becomes more mentally tiring

to be constantly trying to make sense of all these sounds. Also, Norway is a country that adopts new technology and moves forwards very quickly, but at the same time is very keen to preserve various things that are culturally important. Things like being able to go to the mountains or the forests for example and *not* be able to hear any artificial noise, but to preserve that peacefulness in the sound is very important.

Anonymous: Norway

As the title of this book implies, how the world *listens* is of course the focus, but the idea of also exploring how the world *sounds* was admittedly rather compelling. It stands to reason that the unique features and qualities of a place would determine an equally unique soundscape. Addressing this directly within the interviews did have some difficulty, however. Describing how a particular place sounds unique is of course a relative question, requiring the contributor to draw comparisons and identify differences between two or more locales. Not everyone spends significant periods of time between two places. Of those that do, few consciously appraise the different soundscapes and it wasn't appropriate to expect contributors to draw such comparisons, though several respondents did just that.

Voice and language

Matters of prior experience throughout a person's history, driving their present expectations, also emerged within the context of the human voice. Several contributors, particularly those who identified as bi- or multi-lingual, felt that the voice was an important talking point when asked about cultural influences on human perceptions of sound.

The sound of a human voice can develop very personal meaning. I regularly listen to recitations of the Qur'an, but I find that only specific people have developed a voice where I feel that they have truly mastered the Arabic. These are voices that I want to listen to again and again. It becomes important to listen to that specific person deliver those words. I can sing *Dancing Queen*, but it's never going to be the same as if sung by ABBA. That sound from a specific source has a very particular association that you attribute meaning and value to. I'm also quite old-school in some ways of thinking, relevant to sound. If a man introduces himself to me, I will often instinctively consider the depth of his voice in relation to mine. I don't dwell on this for more than a moment but it's interesting to me that I do this.

Hatim: Pakistan

Though Hatim may have felt his perspective on vocal tone to be "old-school", it is true to say that several empirical studies support such an attitude, at the very least in pragmatic terms. A 2016 experiment by Tsantani

and colleagues revealed low-pitched male voices to be significantly pref-
erential, rated more favourably than high-pitched male voices in terms of
trustworthiness and dominance. This preference was asserted by the authors
to be independent of context and therefore applicable to any given situation. The same study also revealed higher ratings of trustworthiness, but
not dominance, for lower-pitched female voices and suggested that preference towards higher- or lower-pitched vocal tone in female voices is more
situation-dependent.

> English, or more specifically the part of the English community that I
> typically interact with, is generally university educated, and I find that
> they speak in a very particular way most of the time – not as expressive
> with their hands or their tone of voice, but more neutral. I was told that
> this way of talking in England is more associated with high society but
> in Jordan, we don't really have that kind of class distinction as commu-
> nicated through sound. I have a UK colleague who once said they felt
> that they sounded ignorant, just because they had a northern accent. That
> made me think if there was any kind of social division based on vocal
> sound in Jordan and I don't think that there is.
>
> Dana: Jordan

More broadly, cultural factors affecting vocal tone and how it is perceived
features in many instances throughout the academic literature. In a 2017
review of patient-clinician communication, Lorié and colleagues revealed
vocal tone to be one of several non-verbal cues that can be interpreted very
differently by patients depending upon their culture, leading the authors to
strongly advise that healthcare professionals should be trained in culturally
specific non-verbal communication. An experiment by Ishii and colleagues
(2003) compared comprehension of emotional words between American
and Japanese participants. They used a form of Stroop test, which broadly
describes the presentation of incongruent stimuli to assess cognitive interference. The classic example of this is the colour/word test in which text
denoting colours is itself presented in colours, some congruent (e.g. 'blue'
in blue letters), and some incongruent ('blue' in green letters). In Ishii and
colleagues' experiment, various evocative words were uttered that broadly fit
into either 'pleasant' (warm, pretty, natural) or 'unpleasant' (tasteless, anxiety) groups. Whether the vocal tone corresponded or not to the pleasant or
unpleasant meaning of the word was varied and the participants were tasked
with identifying each complete utterance as either pleasant or unpleasant.
The results showed that Americans were significantly more likely to judge
an utterance based on the word and would more commonly ignore the vocal
tone in their assessment. By contrast, the Japanese participants demonstrated
the opposite, mostly judging the utterance in correspondence to the vocal
tone whilst ignoring the word.

Tone of voice is definitely an important cultural feature of sound. Language has changed over the years. There's been a reduction in Received Pronunciation and more regional accents and dialects can be heard on the television and the radio. Growing up, my mother discouraged me from developing a local Portsmouth accent. The truth is that it this helped me in business. Working in fashion, a lot of my clients were 'of a certain status'. If I were to accompany them to Bond Street and I sounded 'different' with a regional voice, I would certainly be treated differently. Nouveau riche has changed things to some extent, with people from a wider range of backgrounds coming into money, making it more likely to hear a rich person speaking in a regional voicing. I get a sense that positive reception to regional accents can be something of a 'flavour of the month' however, and that Received Pronunciation is more likely to be well-received long term.

Alyson: UK

Cultural reception to accent is most certainly something that has been explored within academic study. British Received Pronunciation (RP) enjoys higher status as a "socially valued accent" (Agha 2003: p.231), to the extent that even members of the British public who do not speak it would largely (and unfairly) describe it as the 'baseline' against which all other English accents are "deviants" (p.232). Further in line with Alyson's observations, various pieces of research also support the notion of accents changing over time (Hannisdal 2007; Yan et al. 2007). In 1994, Wells confronted the question of whether RP was undergoing a process of 'cockneyfication'. He deduced that it was not, but acknowledged it was undergoing some acoustic change. Observing that the annual British Christmas address, delivered by HRH Queen Elizabeth II, provides a reliable means of examining pronunciation over a significant period, Harrington and colleagues (2000) did just that in efforts to reveal how RP had developed. Taking a total of nine Christmas addresses, first broadcast between 1952 and 1988, their research revealed significant acoustic changes in the Queen's accent. Harrington and colleagues described the shift as being away from what is known as 'conservative RP' and towards 'mainstream RP'. Sometimes referred to as 'English of the older generation' or 'upper-crust RP', conservative RP is commonly characterised by a /j/ sound in words such as *tune* (/tju:n/) and *home* (/hj:Ohm/) and observable in plenty of BBC and ITV period dramas. Mainstream RP by comparison maintains certain 'clipped' features but is arguably a more 'pragmatic' representation of the given letters (cat = kæt), compared to the more ornate quality of conservative RP (cat = ket). If you'd like a listen, there are a few good videos online that demonstrate various RP forms[3].

In addition to how vocal tone can change over time, research has also revealed several ways in which cultural shifts, within what might be otherwise described as a single cultural demographic, can also change our relationship with the voice. For example, a 2001 article by Milroy reinforces Alyson's observation that regional accents are becoming more prevalent in circumstances that were previously reserved for RP. Television presenters are a solid example of this. As a South Yorkshire native, however, I am obligated to press upon you the further observation that this diversification is largely described as an increase in the variety of *southern* English accents. Milroy further posits that the cultural reception of RP has dramatically shifted, due to a substantial decline in the now antiquated belief that RP could be used to identify a speaker's level of education.

Although proving motivation is tricky, as one does not simply ask the Queen why her voice is changing, the assertion made by Harrington and colleagues (2000) is that the primary rational for this vocal shift was a desire to increase the clarity of the message for as many people as possible. Whatever the truth in this instance, the position of research within this area strongly advocates a relationship between the sound of the human voice and broader culture. Assumptions, expectations, and judgements are all intertwined and bound to our interactions with both our own voices, and those of the people in our lives. Have you ever found yourself using a modification of your voice in certain company, then found yourself in a situation with a group of people, some of whom only know you by your natural tone whilst others have only heard your modified 'telephone' voice? If you haven't, I don't recommend it. It's extremely embarrassing.

Religion and ceremony

How sound features in the everyday practice and experience of religion became a topic of conversation that we would return to in many of the interviews. It was often raised without the need for any prompting from myself. With regards to research on the subject, music is of course a significant feature of discussions and debates (Engelhardt 2011). This is particularly true in matters of religious events and their sociocultural implications. As Sykes (2015) observes, there is a long and controversial history in the use of drumming during the Hindu Thaipusam festival in Singapore. Beyond music, a 2012 review by Hackett asserts that the role of sound within religion is undervalued, likely linked to the generalised privilege of visual aspects, particularly in Western religions. This is the case even though certain sounds are emblematic of specific religious practices (from the prayer bowl of Buddhism to the church organ of Christianity), and despite sound being established as a pivotal component of transcendental ceremony and healing rituals. Hackett further observes that, just as sight is privileged over sound, vocalised expressions of sound (such as recitations and hymns) are privileged over other auditory forms. This issue remains underexplored, though relatively recent research has begun to make progress by applying acoustic archaeology to the study of religious practice in order to better understand the relationships between the experience of religion and soundscapes, specifically, as affected by the spatial and material properties of religious buildings (see Kolar 2018). Hackett's (2012) review also provides us with an ideal introduction to works that broke new ground. These include Schmidt's (2002) observations on the participatory and performative value of sound in religious ritual, whether full of joy or full of sorrow, and his insistence that non-verbal expressions (sobs, sighs, laughter, and groans, for example) are key elements of religious experience. Hackett also cites the work of Tuzin and colleagues (1984), whose work explores the physiological sensations evoked by sound and their associations to spiritual and supernatural presences (Hackett references the distinctive whir of the Bullroarer[4], a device prominently used in Australian Aboriginal ceremonies as well as in Māori, Dogon, and native South American cultures).

> Sound is an important feature of the Muslim Call to Prayer. Although it can be loud, it is very melodic and musical. This can be very soothing and relaxing. It puts you in the mood to practice what it is calling you to do, which is to attend the mosque and to pray. When I'm in the mosque, the sermon can be delivered in English, but could alternatively be in Bengali or another language that not everyone attending can understand. In that circumstance, the sound within the mosque can feel disconnected and even flat. But although the Adhan is in Arabic, most recitations from the Qur'an have a melody to them, making it almost musical but also much more recognisable and easier to interpret meaning from. Also, before prayer it is important for Muslims to be clean and so we have a ritual of ablution. This means you will clearly hear the splashing of water as people wash their hands and faces. In Portsmouth, we don't have any purpose-built mosques so I can't comment on how their design would affect the sound. The mosque I attend used to be a theatre, so in terms of acoustics, well, you have theatre acoustics. The sound does bounce around a lot in there, you can feel a noticeable echo. It's also typically just open space, without chairs or other furniture, so this further adds to the echo. Other than recitations from the Qur'an, music is something of a point of contention in the Islamic community. Some scholars believe that music is forbidden, whilst others argue against that. So, in different Muslim households, music may feature very differently.
>
> Jahangir: Bangladesh, UK

> In church the space is so hollow. If a child makes one small sound it will echo all around. Also, there's this expectation for children to be quiet, which I think leads to most children essentially 'holding it in'. Every time I see a child in church, I'm convinced that they're going to belt out a scream in the middle of the service. During the pandemic, the church services did not include singing and that had a profound effect on the overall experience. Some churches would not include any music, whilst others just allowed instruments. Hymns were still present, but you had to internalise the words. For some people this was difficult, particularly

those who are quite musical and enjoy singing. For some, the act of singing is an important way in which they worship.

In terms of cultural difference, one thing I notice in the Catholic Church is that much of a service is the same each time you experience it. The church uses consistent readings, and a service must always include certain points. The exception to this consistency is in the sound, specifically the musicality, which can vary greatly. So, in a church service in Kenya, it is much more likely that there will be drums whereas in the UK it's far more likely that you'll hear an organ. In Kenya, people will dance to the music as they come into the church. The architecture of churches is largely similar between the UK and Kenya, so the acoustics in the spaces are not too different, but I feel that the church experience in Kenya is generally louder. The people are more likely to project their voices and I've noticed a larger proportion of younger people in congregations in Kenya. I think that they are more likely to project and raise the volume. I would describe the UK church sound as more sombre. I actually remember a priest from Africa visiting the UK and delivering a sermon, telling the congregation that they needed to be much louder.

Anonymous: Kenya, UK

The Buddhist temple and the Shinto shrine definitely have their own sound. In terms of architecture, the temple will feature very thick wood and tatami mats on the floor. This absorbs a lot of the sound and creates a more isolated effect. Inside the temple you may hear the sounds of bells and prayer bowls. The sound is quite profound and long-lasting. It's a constant, deep sound that is symbolic of a Buddhist temple for me. At the temple, people will be very quiet. At a Shinto shrine, they use sounds that are more of a high pitch. You will hear vocal sounds of prayer and bells. These are very small bells, sometimes in a series on a stick. These sounds are much lighter. The architecture is also very different because the shrine is open-air, so the sounds are more integrated with nature. The people at the Shinto shrine will often make more sound than at the temple. Part of the ceremony can involve handclapping. I think that the sound of Buddhism emphasises meditation and calm, whilst Shinto can be more uplifting.

We go to the temple at New Year's Eve. The winters are very cold and there will typically be a lot of snow. Inside the temple we will hear the deep sounds, and after the ceremony we will go outside where there is a huge bell. The bell will be hit to mark the New Year, and this makes a very deep and heavy sound.

In Shinto there is great value in respecting nature. I think this does affect our relationship with sound because it encourages us to respect and appreciate sounds that may not be pleasant to everyone. The cicada for example, is very loud, but we would focus on the way the sound is telling of the season, which we appreciate. This is something that many parents wish to teach their children; to not focus on the negative aspects of things, such as loud or distracting sounds, but to appreciate their value, often from the things that those sounds can tell us.

Shoko: Japan

In Sri Lankan culture, religion is often an important element. We'll typically pray every morning and sing around 4–5 hymns in the mornings and in the evenings. In temple you'll hear praying and there's always bell-ringing. At funerals there will be a group of people singing and dancing as they're taking the body to the graveyard. It could be a song or instrumental music with mainly drums for example. The intention is to make people aware that the event is happening but also to give a kind of sound-based impression of that person. The kind of songs or musical themes you'll hear are specific to funerals and you'll often hear the same song at different funerals, but the arrangement of the song is personalised. The core of the piece stays the same, but the instrumentation, tempo, or energy of the music could be more individual.

Laksha: Sri Lanka

The Call to Prayer in Amman is very distinct because of the topography and architecture. Of course, every mosque in the city will present the Adhan, but the cliffs and the buildings echo the calls to make a very large, loud sound all through the city. Historically, up to the early 2000s, the Adhan was a more individual sound for each mosque that would start and end at slightly different times, but the echo meant that all these different calls were crashing into each other to make this huge mess of sound. Now the government has made it so that all the mosques play the same call at the same time, so it is still very loud but much clearer and more recognisable.

Dana: Jordan

There is quite a bit of overlap between the above responses and the research. Comments do emphasise musical and vocal aspects of religious sound, but our contributors reliably went beyond these two aspects. They identify music as a key factor in helping the comprehension and interpretation of religious sermons but also in making them more engaging. They explain how music can facilitate expression for members of a congregation and can be a near-essential interaction within the overall experience of attending a religious event. Across most depictions outlined by contributors, performative music punctuates the majority of ceremonies, processions, and practices, but of course, the features and qualities of these sounds can be very different indeed.

These discussions demonstrated an impressive awareness regarding seldom-discussed sound objects, such as the splashing of water in the ritual of ablution. Exploration of acoustics was also a common feature, addressing matters such as acoustic variation between religious buildings, the effects of such buildings not being built for purpose, and how differences in building materials, props, the configuration of space, and practice-specific objects and events all contribute to the production and shaping of sounds that, for many, are emblematic of their religion. How sound, religion, and wider culture were connected was also referenced in many cases. Examples included: how religious positions on music could have implications for the soundscape of a family's home; how features such as demographics and attitudes within a congregation can be a significant source of sound variation, even within the same religion; and how geographical and topological factors can affect the perception of religious sound and even instigate changes in government policy due to its societal effects.

Chapter 4 summary: "not something I normally think about"

Asking our contributors their thoughts on sound and culture raised a diverse selection of matters for further discussion. Topics included religion, vocal tone, local history, changes in the season, and more. In line with much of the research on the subject, culture meant a lot of different things to different people, and whilst few would assert that any of the topics discussed above are outside the bounds of cultural effects, what first comes to mind when asked about sound and culture, and what matters are deemed most prominent, can all vary between us.

One final observation to make before we close this chapter was the tendency amongst contributors to express a low level of self-rated awareness regarding sound and culture. This is not to say that the interviews did not yield numerous insights, as can be clearly seen throughout this chapter, but it was surprising the frequency with which interviewees would exclaim that they had never considered cultural factors before now and that they were effectively thinking out loud in their responses. I must admit that I do feel the same way. How culture affected my relationship with sound was not something I normally thought about. This made me wonder what the implications of doing so might be.

Notes

1 Euromonitor International infographic on coffee and tea consumption: https://www.euromonitor.com/article/the-worlds-biggest-coffee-and-tea-drinkers (accessed 31.01.2022).
2 A couple of examples of online video compilations of sirens from around the world: https://www.youtube.com/watch?v=CuduG89DrLE; https://www.youtube.com/watch?v=TjBnDsTOGug (accessed 01.02.2022).
3 Examples of online videos demonstrating alternative forms of received pronunciation: https://www.youtube.com/watch?v=g0qShxkuS7Q; https://www.youtube.com/watch?v=1KaXTasMTwo (accessed 01.02.2022).
4 Video demonstration of a Bullroarer: https://www.youtube.com/watch?v=vscBpbao7Os (accessed 06.02.2022).

References

Baron-Cohen, S. (1996). Is There a Normal Phase of Synaesthesia in Development? *Psyche*, 2(27), 223–228.

Birch, H. G., Belmont, I., & Karp, E. (1965). Social Differences in Auditory Perception. *Perceptual and Motor Skills*, 20(3), 861–870.

Breger, I. (1971). A Cross-Cultural Study of Auditory Perception. *Journal of General Psychology*, 85, 315.

Cao, L., & Gross, J. (2015). Cultural Differences in Perceiving Sounds Generated by Others: Self Matters. *Frontiers in Psychology*, 6, 1865.

Carmel, D., Yeshurun, A., & Moshe, Y. (2017). Detection of Alarm Sounds in Noisy Environments. In: Maragos, P. and Theodoridis, S. (Eds.) *25th European Signal Processing Conference (EUSIPCO)* (pp. 1839–1843). New York: IEEE.

Carter, D., & Muller, P. (2015). *The Economic and Cultural Value of Live Music in Australia, 2014*. University of Tasmania, Australian Live Music Office, South Australian government. City of Melbourne and Live Music Office (Australia).

Chua, H. F., Boland, J. E., & Nisbett, R. E. (2005). Cultural Variation in Eye Movements During Scene Perception. *Proceedings of the National Academy of Sciences*, 102(35), 12629–12633.

Engelhardt, J. (2012). Music, Sound, and Religion. In: Clayton, M. and Herbert, T. (Eds.) *The Cultural Study of Music* (pp. 321–329). New York: Routledge.

Fonseca, N. (2014). Soundscapes and the Temporality of Auditory Experience. In: Castro, R. (Ed.) *Proceedings of Invisible Places/Sounding Cities. Sound Urbanism and Sense of Place*. 18th-20th July 2014. Jardins Efemeros. Viseu, Portugal. pp. 523–534.

Ghahraman, M. A., Farahani, S., & Tavanai, E. (2021). A Comprehensive Review of the Effects of Caffeine on the Auditory and Vestibular Systems. *Nutritional Neuroscience*, 2021(1), 1–14.

Hackett, R. I. (2012). Sound, Music, and the Study of Religion. *Temenos-Nordic Journal of Comparative Religion*, 48(1), 11–27.

Hannisdal, B. R. (2007). *Variability and Change in Received Pronunciation: A Study of Six Phonological Variables in the Speech of Television Newsreaders*. The University of Bergen.

Heider, F. (2013/1958). *The Psychology of Interpersonal Relations*. Psychology Press.

Herndon, M., & McLeod, N. (1981). *Music as Culture*. Norwood.

Homan, S. (2019). Lockout' Laws or 'Rock Out'laws? Governing Sydney's Night-Time Economy and Implications for the 'Music City'. *International Journal of Cultural Policy*, 25(4), 500–514.

Ishii, K., Reyes, J. A., & Kitayama, S. (2003). Spontaneous Attention to Word Content Versus Emotional Tone: Differences Among Three Cultures. *Psychological Science*, 14(1), 39–46.

Iversen, J. R., Patel, A. D., & Ohgushi, K. (2008). Perception of Rhythmic Grouping Depends on Auditory Experience. *The Journal of the Acoustical Society of America*, 124(4), 2263–2271.

Kadosh, R. C., Henik, A., & Walsh, V. (2009). Synaesthesia: Learned or Lost? *Developmental Science*, 12(3), 484–491.

Knöferle, K. (2012). Using Customer Insights to Improve Product Sound Design. *Marketing Review St. Gallen*, 29(2), 47–53.

Kolar, M. A. (2018). Archaeoacoustics: Re-sounding Material Culture. *Acoustics Today*, 14(4), 28–37.

Lankford, R. D. (2013). *Sleigh Rides, Jingle Bells, & Silent Nights: A Cultural History of American Christmas Songs.* University of Florida Press.

Lorié, Á., Reinero, D. A., Phillips, M., Zhang, L., & Riess, H. (2017). Culture and Nonverbal Expressions of Empathy in Clinical Settings: A Systematic Review. *Patient Education and Counseling,* 100(3), 411–424.

Maruna, S., & Mann, R. E. (2006). A Fundamental Attribution Error? Rethinking Cognitive Distortions. *Legal and Criminological Psychology,* 11(2), 155–177.

Merckelbach, H., & van de Ven, V. (2001). Another White Christmas: Fantasy Proneness and Reports of 'Hallucinatory Experiences' in Undergraduate Students. *Journal of Behavior Therapy and Experimental Psychiatry,* 32(3), 137–144.

Milroy, J. (2001). Received Pronunciation: who "Receives" It and How Long Will It Be "Received"? *Studia Anglica Posnaniensia: International Review of English Studies,* 2001(1), 15–33.

Moore, O. K. (1952). Nominal Definitions of 'Culture'. *Philosophy of Science,* 19(4), 245–256.

Morrison, M., Gan, S., Dubelaar, C., & Oppewal, H. (2011). In-store Music and Aroma Influences on Shopper Behavior and Satisfaction. *Journal of Business Research,* 64, 558–564.

Ong, J. S., Hwang, L. D., Zhong, V. W.,… & Cornelis, M. C. (2018). Understanding the Role of Bitter Taste Perception in Coffee, Tea and Alcohol Consumption Through Mendelian Randomization. *Scientific Reports,* 8(1), 1–8.

Shigehisa, T. (1979). Intersensory Facilitation of Visual and Auditory Perception in Relation to Cultural Factors. *Japanese Psychological Research,* 21(2), 78–87.

Smilek, D., Dixon, M. J., Cudahy, C., & Merikle, P. M. (2001). Synaesthetic Photisms Influence Visual Perception. *Journal of Cognitive Neuroscience,* 13(7), 930–936.

Spence, C., & Carvalho, F. M. (2020). The Coffee Drinking Experience: Product Extrinsic (Atmospheric) Influences on Taste and Choice. *Food Quality and Preference,* 80, 103802.

Sykes, J. (2015). Sound Studies, Religion and Urban Space: Tamil Music and the Ethical Life in Singapore. *Ethnomusicology Forum,* 24(3), 380–413.

Tsantani, M. S., Belin, P., Paterson, H. M., & McAleer, P. (2016). Low Vocal Pitch Preference Drives First Impressions Irrespective of Context in Male Voices but Not in Female Voices. *Perception,* 45(8), 946–963.

Tuzin, D., Blacking, J., Gewertz, D., de Carvalho, J. J., Kaplinski, J., Kingsbury, H., … & Young, M. W. (1984). Miraculous Voices: The Auditory Experience of Numinous Objects [and Comments and Replies]. *Current Anthropology,* 25(5), 579–596.

Walker, R. (1987). The Effects of Culture, Environment, Age, and Musical Training on Choices of Visual Metaphors for Sound. *Perception & Psychophysics,* 42(5), 491–502.

Ward, J. (2019). Synaesthesia: A Distinct Entity That Is an Emergent Feature of Adaptive Neurocognitive Differences. *Philosophical Transactions of the Royal Society B,* 374(1787), 20180351.

Wells, J. C. (1994). The Cockneyfication of RP. In: Gunnel Melchers and Nils-Lennart Johannesson (Eds.) *Nonstandard varieties of language.* Papers from the Stockholm Symposium, 11-13 April 1991. Stockholm: Almqvist & Wiksell International.

Wong, P. C., Chan, A. H., & Margulis, E. H. (2012). Effects of Mono-and Bicultural Experiences on Auditory Perception. *Annals of the New York Academy of Sciences,* 1252(1), 158–162.

Wong, P. C., Ciocca, V., Chan, A. H., Ha, L. Y., Tan, L. H., & Peretz, I. (2012). Effects of Culture on Musical Pitch Perception. *PloS One*, 7(4), e33424.

Wong, P. C., Roy, A. K., & Margulis, E. H. (2009). Bimusicalism: The Implicit Dual Enculturation of Cognitive and Affective Systems. *Music Perception*, 27(2), 81–88.

Wood, J. (2019). These are the countries where most adults still don't have a smartphone. World Economic Forum. Online article: https://www.weforum.org/agenda/2019/02/two-thirds-of-the-world-s-adults-still-don-t-have-a-smartphone/ (accessed 01.02.2022).

Yan, Q., Vaseghi, S., Rentzos, D., & Ho, C. H. (2007). Analysis and Synthesis of Formant Spaces of British, Australian, and American Accents. *IEEE Transactions on Audio, Speech, and Language Processing*, 15(2), 676–689.

Yu, L., & Kang, J. (2010). Factors Influencing the Sound Preference in Urban Open Spaces. *Applied Acoustics*, 71(7), 622–633.

5　Sound and place

Wightlink Ferry Terminal, Old Portsmouth. 27th of October 2021. Early evening

I'm standing next to the mouth of the Solent. There is less human sound here but also less cover from the elements and the wind becomes a prominent feature of the soundscape. In the distance behind me, al fresco diners clink their cutlery, their accompanying voices upbeat despite the cold. I'm hungry. Continuing along the parameter and I chance upon a lesser-experienced sound. Directly ahead of me the car ferry from Portsmouth to the Isle of Wight is pulling out of port. The ferry itself towers above me. It moves slowly. The sound is deep and steady. It feels measured and subtle. The low frequencies are permeating but relaxing. Then the ferry pulls to the left and I get to play the game of 'how many different car alarms can you hear?'. I can hear five.

How does the world sound?

How the World Listens is, first and foremost, a book about auditory perception – about our relationship with sound, how we make sense of it, extract information from it and conduct our lives relative to it. The title may not be *How the World Sounds*, though I must admit to finding that topic a compelling one. It most certainly stands to reason that the diversity in general between locales around the world would dictate that the soundscapes of these places would be just as varied. This assumption is further reinforced by the very existence of what is often referred to as 'sound tourism' – the visiting of different places specifically to experience a unique sonic character. As Sebastian Bernat observes: "Sound helps one to understand a particular place […], it enlivens a particular space, fosters relaxation, and strengthens or reduces aesthetic sensations. Sound impacts the quality of landscape and shapes its character" (2014: p.109). Based primarily on the comments of our contributors, this chapter embarks on its own form of sound tourism and considers the relationship between sound and place.

DOI: 10.4324/9781003178705-6

Conceptualising the soundscape

In 2010, Herranz-Pascual and colleagues put forward a detailed conceptual framework on soundscapes, specifically for the methodological function of studying them more effectively. Their model identifies person, place, and activity as the three primary determinants of a soundscape. Person, which can be applied to either an individual or a community, collates together factors that include health, social demographic, lifestyle, and personality. Place describes aesthetic, acoustic, and interactive features that are themselves a grouping of factors such as climate, topography (the arrangement of natural and artificial features of an area), pollution/environmental quality, functions of place (home, work, retail, social, etc.), and safety. These factors can affect our experience of a soundscape by way of our activity as a mediator, but also directly, evoking user-experience dynamics that include previous experience, satisfaction, familiarity, expectation, identity, and information. As with the majority of conceptual frameworks, a picture paints a thousand words and so I would recommend taking a look at the Herranz-Pascual and colleagues' own illustration[1]. However, the above does provide us with a detailed checklist of elements to look out for in our interviews.

The following year, Pijanowski and colleagues (2011) coined the term 'soundscape ecology'. This new idea was presented in what is arguably a very concise and comprehensive conceptual framework for understanding sound and place. In a nutshell, Pijanowski and colleagues' model[2] centres upon the soundscape as a collective of space, time, interaction, and composition. Feeding directly into the soundscape are influences from the natural environment (comprising biological and geophysical elements that include landform, wind and water, biodiversity, and habitats) but also the built environment (human-made structures and land usages such as roads, building, artificial waterways, urbanisation factors, and agriculture). The processes of the natural and built environments influencing the soundscape are referred to as *geophony* and *anthrophony* respectively. Another hierarchical layer is also identified that impacts the natural and built factors of soundscape, with atmospheric effects (weather, climate, temperature, seasons) influencing geophony and the human system (policies, values, needs, and behaviour) affecting anthrophony. Our exposure to the soundscape means that it feeds back into the human system, influencing our attitudes and perspectives that will in turn manipulate how we further develop our built environment. Non-human exposure to the soundscape generates a further feedback loop with the natural environment, potentially altering animal behaviour which, in turn, can indirectly impact upon habitat and biodiversity. With so many factors influencing a local soundscape, some in potentially dramatic ways, it seems highly unlikely that soundscapes, themselves an emergent reflection of a complex interrelation of factors, do not possess unique sonic features and qualities – but what effect do these differences have on people's everyday lives?

Continental soundscapes

Typing the relevant terms into the standard Google search bar, "[your continent of choice] AND soundscape" reveals enough diversity in the results for each continent to indicate not only that places around the world sound objectively different in acoustic terms, but also that our sense of meaning and value about sound differs around the world. For instance, "Africa AND soundscape" first returns video links with a strong emphasis on nature, rural soundscapes, and relaxing ambiences. Various subsequent weblinks interpret our search more in terms of music, with numerous references to tribal drumming, chants, and associations between music, culture, politics, and other forms of art. Alternatively, "Asia AND soundscape" highlights scholarly results first, with links to review articles and references to several soundscape appraisal projects. Environmental concerns associated with urban living are observable in several webpages, whilst the links to video content emphasise river and rainforest soundscapes. In the returns for "Europe AND soundscape", the prominence is very much on policymaking and legislation, public health, and environmental concern. "America AND soundscape" by comparison, presents us with a pretty balanced and wide-ranging set of responses that include matters on architectural acoustics, health concerns over noise pollution, music, national history, issues of cultural identity, links to soundscape-derived artistic works, and even articles on in-car entertainment.

Documented in what is arguably one of *the* seminal texts on the subject, R.M. Schafer and colleagues (1977) from the World Soundscape Project at Simon Fraser University visited numerous cities and other locales between February and June of 1975. Amongst others, their tour comprised the cities of Paris, Amsterdam, Stockholm, Stuttgart, Vienna, and London. The researchers all piled together in a Volkswagen campervan (the story at this point sounding more like the synopsis for a charming independent film than a research expedition) and kept individual sound diaries. Schafer took responsibility for pooling the writings together and selecting the passages that would form the account of their experience. Vancouver served as their baseline for comparison. On this odyssey was observed differences, unique features, and qualities of sound: traffic noise and public transport; church sounds such as bell ringing, different types of organs, and varying loudspeaker arrangements for sermons; social venues including bars, clubs, restaurants, coffee shops, cafes, and public houses; telephone mannerisms; various acoustic effects of the weather, particularly rain and snow; retailers and street sellers; industrial and factory noise; even public sanitation, sewerage and drains. Their work paints highly discrete sound pictures of these different locales, strongly reinforcing the notion that the world soundscape is indeed full of dramatic difference, depending on precisely where you are.

The Big Ice

Sadly, I was unable to recruit any contributors with whom I could discuss the continent of Antarctica. However, my curiosity did drive me to have a little rummage online and I was quite intrigued by the results. Soundscape studies of the Big Ice are relatively recent, and motivation for them was clearly spurred on by ecological concerns and to inform on climate change. As Philpott and Leane (2021) reveal, the unique soundscape and wider ecological and topological nature of the Antarctic demands that sound take priority over image and that human hearing has been shown to adapt uniquely to this environment over time. Though an important factor, this is not simply due to the highly reduced visibility across the Antarctic regions, but also because understanding the sounds of the ice and the distinct wildlife is essential to survival for all who dwell there, humans included. Philpott, in particular, is a recurring name within studies of the Antarctician soundscape. Her works examine, amongst others, the history of music and 'sledge songs', providing entertainment and comfort on gruelling Antarctic expeditions (Philpott 2013), and the application of the unique sonic characteristics and atmospheres of the continent to the composition of sound art (Philpott & Samartzis 2017). For Philpott, the soundscape of the Antarctic is a compilation of geophysical sounds emerging from the landform and the movements of the wind and water. The deep and sustaining cracking and calving of the ice. The howling of the blizzards and the crash of the waves. This is punctuated by the biological, as penguins and seals jostle about the space, but also the less common, yet hugely distinct clanks, scrapes, and shuffles of human activity.

Application of soundscape studies

In her 2019 doctoral thesis, criminologist Katherine Herrity applied aural ethnography methods (analogous to soundscape studies) to a local prison, presumably in the Leicestershire region of the UK (the prison itself was anonymised to "HMP Midtown"). Herrity's investigation revealed numerous insights into the dynamics of a UK prison, exploring various matters of power, order, and emotion. Prison officers jangled their keys performatively as an expression of authority (or simply through nervous habit), with the capacity to emotionally destabilise prisoners. Prisoners developed enhanced functional dependency on sound, with the architectural nature of prisons obscuring visuals so greatly, sound afforded a much greater perception of space and situational awareness.

HMP Midtown is just one of many ways in which soundscape studies can improve our understanding of a subject. Soundscape studies have proven profitable approaches to research in several domains. These include travel and tourism (Liu et al. 2018), environmental impact assessment (Duarte et al. 2021), architectural design (Schulte-Fortkamp & Jordan 2016), and urban planning (Aletta et al. 2016). Our reliance upon visual stimuli arguably translates over to research. If we want to understand something better, the first instinct is nearly always to look at it more closely, but what are the missed opportunities to our understanding when we only look, but we don't listen? Central to the ambition for this book, Ari Kelman (2010), asserts that studying a soundscape can reveal insight into the *human relationship with sound* and reveals to us the social production of meaning.

Some local perspectives

The original structure for this book was actually intended to be something of a globetrotting adventure, exploring our relationship with sound on a continent by contentment basis. This did, however, prove itself to be deeply problematic for two primary reasons. The first reason was the geographical distribution of contributors' comments on specific issues was too uneven to consistently draw fair and balanced conclusions between particular parts of the world. This was potentially a result of my own failing, as a greater sample size or substantially extending the length of each interview may have overcome this issue. But, in the midst of the interviews themselves, sticking doggedly as I did to the ethnographic ideals of the methodology (namely, allowing the interviewee to set the agenda), it was clear that consistency in topics covered could not be standardised by place. Simply put, different people wished to discuss different themes, and so it was not possible to meaningfully compare one part of the world against another. I bring up this admission of structural change, not because I'm masochistically desperate to reveal my own failings, but because this particular section is the exception to the above rule precisely because its theme *is* place. That said, the intention here is not to prove the difference in soundscape between regions of the world, but rather to explore sound and place with a diverse range of people, and to reveal a wide range of themes relevant to the human relationship with sound for further exploration. The following discussion is therefore structured on that originally intended continent by continent basis (well, kind of – it's more of a convenient land mass by convenient land mass separation), starting with Europe before progressing to Asia, the Middle East, and Oceania, then on to the Americas and finally Africa. As you might expect from opportunistic sampling, in which the availability of participants largely determines the sample, the place being explored by several contributors was the humble United Kingdom. We start our journey there.

A question of sport

> Here in Portsmouth the sound I can always hear, and all my neighbours can recognise, is from the local football ground in the afternoon when Pompey play. You can always hear the fans singing, which first tells you that it's almost 3 o'clock, we're going to kick off in a minute. The sound is very emotional. If it gets to around 5 o'clock and the streets sound rather quiet, you know that Pompey lost. During the game, if we're sitting out in the garden, you can hear the emotion in the crowd, it's a wonderful sound. At the same time though, the sound from the football can change my routine at the weekend. Sometimes I'll be getting

ready to go shopping on a Sunday afternoon, step outside, and hear the sounds from the stadium, which will tell me that there will be a flood of fans in the streets soon. I might then decide to leave the shopping for another day.

Si: China, UK

The above notion of utilising the sounds from a sports stadium in order to determine aspects of how you plan your day is a fascinating thought. On reflection, I'm confident that I have used sound in this way at least once before. In terms of academic literature, there is little available that explores such a relationship with sound directly, but research can certainly attest to sound as being a vital, embodying facet of sport, particularly within community behaviour. Deep connections have been evidenced between sound and shared experience, articulation of identity, and reflection amongst fans (Kytö 2011). A 2006 aural ethnography of Scunthorpe United (selection bias, being a mere stone's throw from my old hometown) observed that, despite progressive strategies that sought to create a sense of 'placelessness' in football, supporters resist this, often using song to reinforce and maintain place-related identities (Clark 2006). Crowd noise has been argued to be one of the most crucial factors in the 'home advantage' to the extent that the sound of a crowd can even determine the outcome of a match. This is something that supporters have an acute awareness of, not just in England, but across the world (Marra & Trotta 2019). In a 2006 article on prediction of meaning from crowd noise, Hayne and colleagues put forward a four-factory theory of crowd noise production. They assert that the effort of the individual, the total crowd size, the *synchronicity* (are the members of the crowd all focussed upon something at the same time, or do they engage more sporadically?), and the *diffusion of orientation* (is a singular object/event unifying the whole crowd's attention, or are different members focussed upon different things?) each have dramatic effects on the nature of a crowd's acoustic quality. From this theory, it could be further postulated that if a person received regular exposure to certain crowd sounds within a persistent context, and over a suitable amount of time, they could determine a substantial amount of information on the nature and emotional state of that crowd. Connecting this back to the notion of changing one's routine in response to crowd noise, the impact of football games upon public disorder was certainly handed an unfortunate case study in June of 2021, when England lost the Euro 2020 final and various news outlets published the aftermath upon the streets[3]. I personally remember following four individual England matches within this particular tournament, and I never saw a single player kick a single ball. The local pub a opposite my house provided a chorus of vocalisations that meant every goal, near-miss, questionable tackle, and bad referee decision was discernible, whether I cared to receive this information or not. As each match neared its conclusion and the raucousness-rating of the revellers outside reached fever pitch, I must admit to feeling the compulsion to stay indoors.

The quiet revolution of electric vehicles

With an emphasis on concerns for public safety, the sound of both e-scooters and electric cars has featured prominently in scientific literature, particularly between 2019 and 2022. The general consensus amongst researchers is that, love them or loathe them, e-scooters are not a short-lived trend and will continue to see usage across the world for the foreseeable future, as they meet an immediate demand in public transport (Gössling 2020) and have evidenced benefits to travel time, money-saving and (be still my middle-aged heart...) *fun* (Christoforou et al. 2021). Their low-noise profile actually appears as a key benefit of e-scooters in some studies (e.g. Gebhardt et al. 2021).

> In Portsmouth, one of the things I notice is with these new e-scooters. They don't generate any sound. I think that's very dangerous, especially to older people. My own hearing is not as sharp as it was, and these scooters can travel up to 30–40 miles an hour. You can look around and see no traffic, then you start to cross the road and one of these scooters comes bearing down on you, but you're not hearing anything. I think they should use some kind of artificial sound, just so people can hear them.
>
> Mehrdad: Iran, UK

Articles addressing pedestrian safety do acknowledge Mehrdad's precise concerns regarding the risks of e-scooters, posed by their relative quietness on the roads (Maiti et al. 2019; Sherriff et al. 2021). However, at the time of writing at least, very little has been empirically investigated as to the sound-related risks of e-scooters and there is little reason to believe, genuine as this concern may be, that such a risk will reduce the further proliferation of e-scooters across the globe.

> I think that in the UK, we are on the verge of a 'quiet revolution'. Electric vehicles for example, take away the noise of the combustion engine, but there is still a tension between electric and petrol vehicles in which the former needs to be made louder. The sound of an electric vehicle reversing is actually quite a beautiful sound, it is a 'science fiction whale' of a sound that has been designed to draw the attention. It intrigues me that, in the near future, the ambient sound level, even in a city, will be so low that we can hear bicycles from more than 100 meters away, the birds in the trees, and the quieter sounds of nature.
>
> Ludo: Italy, Egypt, UK

Literature examining low-noise in electric vehicles now has over a decade of research history, suggesting that the e-scooter phenomenon is simply too recent, and that their risks to pedestrian safety will be empirically addressed in the near future. Sound, pedestrians, and vehicles form a complex relationship with lots of conflating factors. The academic literature on pedestrian behaviour suggests that obscuring sound by wearing headphones does reduce situational awareness to the extent that a person is less likely to look for traffic when crossing a road if relevant traffic noise cannot be heard (Wells et al. 2018). Although this appears to not yet have been empirically investigated, research has put forward the idea that sound cues could play a vital role in helping pedestrians navigate vehicles by way of interpreting drivers' manoeuvring intentions (Soares et al. 2021). A study by Faas and Baumann (2021) sought to unpick matters of sound and pedestrian safety in electric vehicles from conflating issues raised by that other near-future ubiquity – autonomous (or self-driving) vehicles. Gathering pedestrian participants together in a car park, the experiment required them to navigate around four iterations of electric vehicles which would be either with or without a simulated engine sound[4] *and* with or without what was called an "external human-machine interface [eHMI]" (p.1). An eHMI describes some means of communicating information concerning the autonomous vehicle to pedestrians in order to support safety and to counter distrust in the technology. It may simply indicate to the pedestrian that the vehicle is in 'self-drive mode', but it could also indicate the manoeuvring intentions of the vehicle. The results of Faas and Baumann's (2021) study revealed that both the eHMI systems and the simulated engine sound were important, as they significantly improved pedestrian comfort and trust ratings towards the autonomous vehicles. Current eHMI designs do occasionally make use of sound cues (e.g. Benderius et al. 2017), but this is not consistent, and, in most cases, sound is supplementary to light-based cues, suggesting that in a future of autonomous, fully-electric vehicles and e-scooters, the "quiet revolution" postulated earlier by Ludo is a likely outcome, and that wearing headphones whilst navigating the streets is only going to become increasingly ill-advised.

Europe and the Russian Federation

Europe has a notable track record in developing regulatory systems to positively influence city soundscapes, putting forward numerous action programmes and policies that address matters of noise pollution as a means towards improving public health and quality of life (Adams et al. 2006). Various European bodies including the European Environment Agency (2014) and the European Parliament and Council (2002) have championed progressive soundscape positions throughout the 1990s but also pretty consistently in recent years (see Kang et al. 2016). The characterful architecture that stretches across the continent has been cited as a crucial factor in its unique sonic profile (Schulte-Fortkamp & Jordan 2016), whilst historical impacts such as the aftermaths of the First and Second World Wars have also had dramatic and wide-ranging consequences for the European soundscape (Tańczuk & Wieczorek 2018). Reviewing the responses from our contributors on the matter of sound and place, we start with a selection of the more concise observations.

I was born in Madrid but spent many of my childhood years on a farm. So, my experience of sound in Spain in that instance was the birds, the wind, and other sounds all around me.

Angélica: Spain

There is a particular bird in Italy, the Eurasian collared dove, that produces a distinct call and really evokes a sense of home.

Anonymous: Italy

Where I live in Russia it is a grey soundscape but with regular bursts of bright and sharp laughter.

Mark: Russian Federation

I've always lived in cities, which are filled to the brim with sound pollution from cars, heavy machinery, construction, and trains. Some of my golden moments with sound are in winter after a heavy snowfall. The snow on the ground and in the air absorbs all that sonic pollution. Suddenly you're walking through very familiar places that sound completely serene. You just hear yourself and the snow that surrounds you.

Rasmus: Denmark, Sweden

St. Petersburg is loud. There is a lot of traffic, emergency services alarms, and sirens. In the summertime there are a lot of street musicians around the city. Many cars will have loud music playing. I would say that Russians generally love listening to very loud music in their cars. The people are themselves, not very loud. We are not talking too much, and not on the streets. There are some people who do love to sing songs when they've had a drink.

Tim: Russian Federation

When I think of sounds that give me a sense of 'home' I think of sparrows. Maybe it's related to some happy memories of being at home in the summer as a child. Standing on the balcony with the countryside in front of me and listening to the birds. When I think of my home in Spain, I think of that memory and that sound of the sparrows is a key part of that memory.

Lucia: Spain

There are also some sounds that are more unique to Serbia, particularly musical sounds. There is the Dodola, which is a ritual chant sung by children during the summer that is meant to bring rain to the village. There are also several distinct regional musical instruments such as the Frula, Gusle, Tamburica, and the Šargija.

Anonymous: Serbia

It is pretty quiet, particularly for Belgium, which is quite a densely populated country. We have quite a few animal sounds, both of our own pets and also around the neighbourhood so you have quite a nice, natural sound scene. You also have the sounds of tractors and other rural sounds. The birds also stand out. We have a lot of Jackdaws.

Brecht: Belgium

If I think of the soundscape in Trondheim a Saturday night, and I think of the difference in comparison to Portsmouth in the UK. Cities in Norway are definitely quieter. There are less vehicles on the road, there isn't the same sort of party culture despite also being a university town. I don't like to stereotype, but I can't help but feel that it's a very British thing to get drunk and loud. This isn't to say that Norwegians never get drunk and loud, but the sound is definitely different. During the day in Trondheim, you will hear children a lot more. Schools and kindergartens take their children outside and around the city and the urban environment a lot more. It feels like Trondheim sounds different in both loudness and pitch. The pitch feels lower. That makes it feel somehow gentler. It's almost more malleable in a way. If I was to hold the sound of the city, it would be soft, mossy, and marshmallow-like. In Portsmouth there is what feels like a constant, high-pitched buzzing around the city. If I was holding that sound, it would be more like a brick or jagged piece of concrete.

Anonymous: Norway

Loudness is certainly a prominent feature in the above descriptions and a focus upon sounds of the streets is also a commonality. Brief references to music, seasonal changes and climate are made, whilst sound objects such as traffic noise and birdsong are recurring features. Contributors identify the traffic, the sirens, revellers in and around social venues, construction, and industrial sounds as *of* the city, compared to the, wind in the trees, the rustling leaves, crackling fireplaces, and chirping insects of the countryside. Birds are prolific sound-makers, featuring more consistently in both urban and rural soundscapes across the world. There is also a strong distinction with regards to descriptive terminology. Cities are loud, heavy, full, rumbling, and lively, whilst the countryside is quiet, serene, numb, silent, and isolated. There is also a distinct absence of clear preference between city and countryside soundscapes, with most respondents avoiding explicit judgements, or presenting a more balanced or contextualised opinion. What I found to be most interesting in the comments is that despite being brief, many contributors quickly move on from detached observations of sound objects or events to describe their own interpretations and feelings – almost as if the limitations of objective description of sound are encouraging them towards more introspective considerations.

Sonic boundaries and the sounds of city and countryside

From the above comments, it would be difficult to suggest that there is no difference in sound between the cities and the countryside. This is broadly true both in terms of sound objects and events, but also in terms of descriptive and perceptual aspects of the soundscape. That said, not all boundaries appear to be defined equally, with some contributors drawing our attention to the ways in which a place can integrate and blur, or separate and isolate, its urban and rural soundscapes.

> We live between Brussels and Antwerp, in what I suppose is a town centre. The immediate space actually feels quite rural, but we are still relatively close to much larger urban centres and there is almost always this constant sound backdrop of something like a highway or our railway, but also the sounds of planes passing overhead. It does create this sense of being mostly in nature but not totally disconnected from urban environments.
>
> Brecht: Belgium

> In Serbia, you have the capital city of Belgrade. In terms of soundscapes, Belgrade is noise, traffic, nightlife. Then you have everything else which is river, birds, the sounds of nature. Of course, there are other towns and areas where people live, but these places are more embedded in nature. There really is a line between the sound of Belgrade and the sound of everywhere else across Serbia.
>
> Anonymous: Serbia

These responses raise several additional opportunities for further commentary, but the most prominent element that emerged for me was in the variations of the urban and country soundscapes that our contributors were describing. The impression is made that the sonic boundaries between city and nature can vary in their distinctiveness, and that different cities across Europe can possess a sound character that would more readily apply to a small town in other countries. Rural sounds are not always confined to rural spaces and otherwise urban environments can be permeated, or accented, by more natural sources. The reverse can also be equally true, with various urban sounds sometimes able to stretch out towards locales that would otherwise be described as rural or suburban.

Meteorological effects

Despite climate featuring within several conceptual frameworks on soundscape, including the two outlined at the start of this chapter (Herranz-Pascual 2010; Pijanowski 2011), the specific impact climate and other meteorological factors can have upon soundscapes is not as well-documented, with the notable exception of its relationship with climate change (Halliday et al. 2019). A small number of studies have documented some observed differences in soundscapes between seasons. Mullet and colleagues (2015), for example, recorded various ambient sounds across the Kenai National Wildlife Refuge in Alaska between the months of December and April. Their results revealed significant spatial (how localised or spread out certain sounds are, and also their precise location within the locale) and temporal (the regularity and commonality of specific sounds) change in the soundscape between winter and spring. In those winter months, geophonic sound (primarily referring to the wind) was greater but predominantly during the night and localised around the lakes. Biophonic (wildlife) sounds decreased in general but also became more concentrated nearer rivers. Lastly, what Mullet and colleagues refer to as *technophonic* (meaning human-made, technological/mechanical or digital) sound also decreased in the winter and become notably less spread out, concentrated within urban areas. This not only makes perfect common sense, but it also resonates with various comments below, in which particular meteorological factors (namely temperature and weather) can affect every aspect of the soundscape, from the sound of wildlife, to those of human activity, to the acoustic qualities of the space itself that thereby colours every soundwave in the environment.

Iceland feels very exotic when compared to a European palette. It has pale blues and pinks, deep greens, and blacks, compared to European browns and dark hues. In terms of sound, Iceland presents a profound silence. There are places where there is simply no sound, such as towards the top of some of Iceland's mountains, in the highlands where you can see the ocean, a huge sky, and there's just not a single sound. No wind, no sense of life, nothing. It's a remarkable experience. In this environment you start hearing yourself. Whilst we use sound in the environment to get a sense of ourselves in space, silence gives us a sense of our body, our self. You can hear the sound of your heart beating, the blood flowing through your body, and a sense of your body physically vibrating.

Nicolas: Iceland

In Greece in August, it is very hot. All the windows and doors are always wide open. The inside and external doors all stay open and so you can hear every little sound from outside. The sound of the cicadas chirping and other insects outside is a real 'summer sound' for me. I love that sound. This makes me think, because it is a very repetitive sound and I'm asking myself why I don't mind.

Anonymous: Greece

In the main street in the towns and cities of Spain, you have all these packed bars and cafes, but the warm temperature means things are very open, so it's really easy for people to run in to each other because there are just so many people. This makes a particular kind of sound. It is very loud, not just because the individual voices are loud, but because there are so many different voices, and because they are so expressive and emotional. The sound of human voices laughing and chatting is a big element of this sound, but it's also chairs being dragged, plates crashing, glasses clinking for 'cheers' – all those things together. The sound of the radio is also very common in Spain. Homes and bars almost always have music playing on the radio throughout the day. If not music, then the news, but the radio is always on. Again, what is really different here is that all of this is outdoors because it is so warm. In the south of Spain, it can be very dry, so the sounds of birds often stands out whilst, in the north, you will hear a lot of running water as there are a lot of creeks and rivers. Of course, if you go to the more mountainous regions, you may hear deer or wild goats. Across Spain we have the beach, we have mountains, we have the deserts, we have forests. It's so different and so the sound environments will be very different too. In the south of Spain, it is customary to have parades during Easter. You'll hear the sounds of an orchestra, trumpets and drums accompanying the parade. You'll also hear the bells of the churches all around. Every region has very different traditions and so practices are different, and the sounds made are different too.

Lucia: Spain

I've been living in the countryside since 2005. Quietness is part of the home. In the summertime the most homelike sound is birds singing and leaves moving in the wind. In the winter, the crackling and humming of the fireplace is the sound which most feels like home. Finland is big country with a lot of nature and city life. Living up here in a small town has changed how I feel about city soundscapes. I dislike the traffic and the city rumble, but I miss the people cheering and their lively talking.

Heikki: Finland

Rounding off our European-based commentaries, we can see further instances of geophonic soundscape factors relevant to climate, both warmer and colder temperatures, but also based on proximity to geographical features that sonically differentiate the creaks, the rivers, to the mountains, the deserts, the beaches, and the plains. Traditions and festivals bring distinct social and cultural dimensions to the soundscape, but also draw people and place together through activity and community interaction that, in turn, contributes further sound to the chorus of the soundscape. Being steeped in tradition, these various dimensions of interaction between person and place bring with them familiarity, expectation, and identity but also establish a framework for feelings and expressions, influencing how people within a place communicate with each other, again affecting the soundscape, this time through their social and recreational vocals.

Asia and Oceania

In the early 2000s, the number of publications on soundscapes began to increase, with dramatic (almost exponential) leaps every five years. As of 2018, publications on soundscapes from Western countries (particularly the United States, the UK, and several other Western European countries) greatly outnumbered those from Asia (especially outside of China and Japan). Whilst the origin of a soundscape study does not always dictate the country of subject (Eastern countries do feature more evenly in this regard), there remained an imbalance between Eastern and Western perspectives on soundscapes (To et al. 2018). This highlights definite gaps in our understanding of the human relationship with sound across the world – with inconsistencies in the global map of place, people, and soundscape.

Defining the Eastern soundscape

In a soundscape exploration of several Indonesian cities, Colombijn (2007) observed that many countries across the East were undergoing significant and rapid modernisation and globalisation, with heavy impact upon the soundscape. Colombijn's research described Indonesia as intensely noise-centric, but what is particularly relevant to our own exploration of the Eastern soundscape is the classification system used to help reveal and organise soundscape components: space, street, modernity, power, and intimacy. Space denoted geography and global position of sound, but also psychoacoustic effects such as urban symbolism (sounds that primarily served as identifiers of the city). Street referred to any sounds one might call 'out in the world' (traffic, car horns, trains and buses, shops and street-based vendors, musicians, etc.) whilst modernity denoted sounds that imply technological progress (smartphone ringtones and interface sounds, the gliding tones of electric vehicles, etc.). Power described sound with an underlying 'show of force' that could include militaristic sounds, rallies, protests and marches or simply motorists revving their engines. Finally, intimacy described sounds of otherwise personal moments that were nevertheless publicly observable (Colombijn uses the example of a newly-married husband and wife leaving a ceremonial building and the furore that surrounded them). With this in mind, we now take a look at a selection from across various Asian, Middle Eastern, and Oceanic countries.

> During most of the national days of celebration, the sound of fireworks has become a very common feature all around the area. In Malaysia we have three prominent cultures: Chinese, Indian, and Malay. Each of these groups has their own set of national celebrations. We have Chinese New Year, Malay New Year, the Diwali season, New Year's Day, Christmas; we have a lot of national days of celebration, and we have a lot of fireworks. What's interesting is that our neighbour, Singapore, has the

very same community makeup but you will very rarely hear the sounds of fireworks. Their laws are particularly strict and for using unlicenced fireworks you could be immediately arrested.

Anonymous: Malaysia

Indonesia has a majority of Islam citizens. There are thousands of mosques spread all over the country. Sound of multiple Adhan from different mosques echoing makes me feel like I am home.

Anonymous: Indonesia

I come from Tehran, which is the capital city twice the size of London. As a big city there is a lot of sound; a lot of hustle and bustle. In the city, you have to maintain an awareness of where you are. Obviously, there's more sound from cars and other sounds, but I think it's the sound of people and their speech that my attention tends to be drawn to.

Mahmood: Iran

You cannot step out in the cities in India without there being inane amounts of traffic. What stands out for me about India is the distinction between natural and city soundscapes. There is a very clear border between them. One kilometre into the city and there is absolutely no sound from nature, and the same distance out into the countryside and there is no sound you can hear from the city.

Anonymous: India

I live in Shenzhen one of the fast-growing cities in China's history. There's construction happening everywhere. In the elevators and public transit, there are usually adverts, that are alive with sound. At the same time, all the taxis and buses, plus a big percentage of the cars are electric, so the noise pollution is quite reasonable. I find that people hear tend to use the car horn less. Every time I cross the river into Hong Kong, I find the traffic noise pollution unbearable after getting used to the sounds in the mainland.

Marc: China

In Christchurch there is a significant difference in that the natural and city sounds feel a lot more overlapping. Out by the river there is a particular sound that it is at once both constant and changing. The sound from the river is this shallow, slow-flowing of water that largely stays the same, but the environment around it changes as you follow along the river. Some areas are more wooded whilst others are more built-up, and the river actually winds its way through the city. What stands out is how these changes to the environment around the river affect the river's sound as you perceive it. You also don't need to move to observe this change, as the particular time of day also had the same effect. I could visit the same point of the river at three separate times during the day and the sound would be different.

Anonymous: New Zealand

In Shanghai there's the alarms, the traffic, a mixture of sounds from the crowds and the music from the shopping malls. These malls play a lot of popular songs and many of them play very loudly. You can hear them from outside as you walk by. Because different shops play different songs, they all mix together. Every day, at around 5–6pm, a group of ladies start to dance in the square with music playing very loudly. Dancing in public is a common occurrence and this happens in many places, not just in Shanghai. Any city in China where there is a large open space you may find groups playing loud music as part of a dance-exercise class.

In the eastern parts of China, the sound of language is very soft. On the coast of China, people get information from the outside world very easily, meaning they can access all kinds of new things from other countries. I think that this can make them more polite and that affects the sound of their voices; it makes them very soft spoken and mild, almost as if they are singing a song. But in my hometown nearer the centre of China for example, the people naturally speak very loudly. It makes others who are not from the town think that the people are very bad tempered. I think that the loudness is something to do with the living style and the environment. People here like spicy food. They live in a mountainous place with a very humid temperature, and most of the people have to work very physically hard. They have to be strong, and I think this all affects their personalities and that the sound of their language shows this kind of strength.

Anonymous: China, Hong Kong

Within the above comments, the variations in distinction between urban and rural soundscapes feature prominently, as they did across Europe. Responses strongly emphasise the street classification of Colombijn's (2007) framework, with intimacy mentioned very briefly alongside matters of acclimatisation and how time of day can factor into the soundscape. Power and modernity both feature in the last of the above responses, the former with regards to personality and vocalisation, and the latter in terms of varying global connectivity. Responses also feature various positive soundscape elements that resonate with the classification of space, such as the melodic echoes of the Adhan, and the various sounds of festivals and celebrations.

Many of the commentaries did feel somewhat matter-of-fact and focussed upon more objective soundscape features than soundscape qualities. A 2016 review by Aletta and colleagues identified eight descriptors of perceived soundscape 'quality' from the body of academic literature: annoyance, pleasantness, tranquillity, music-likeness, affective quality (the extent to which the soundscape evoked positive and/or negative emotional states), restorativeness ("the potential of the environment to re-establish certain cognitive capacities related to human information processing" – Negrin et al. 2017: p.1735), appropriateness (how 'fitting' a sound is in context of its place), and vibrancy (eventfulness, excitement). Looking across the various responses above, what is commonly made clear are the more objective observations regarding sound objects and events. What is less apparent from the text, however, is the underlying qualitative and affective response to the sounds. For example, the anonymised observation that Malaysia hosts a wide array of cultural events could be interpreted as a positive review of the soundscape's vibrancy or a negative description of the annoyance it may instil. In this particular instance, the contributor's emotive tone of expression betrayed their qualitative assessment, which in this instance felt mostly positive. The sounds as described could be classed as having high appropriateness, but there are limits. I also got the impression that in some instances, what one might describe as 'soundscape fatigue' can set in, and what was once exciting and colourful steadily becomes annoying and disruptive.

The commentaries above emphasise the powerful effect of local culture on the soundscape's features and acoustic qualities, but also the effects of context on our auditory perception. Referring back to Herranz-Pascual and colleagues' (2010) soundscape framework, activity mediates person and place, meaning that activity type and activity needs will heavily impact what we want (or don't want) from a soundscape. Its resultant appropriateness (meaning the extent to which the soundscape fulfils those needs) is crucial in our qualitative assessment of the sound.

Different beeps and louder dogs

In Japan the street crossing sounds are different to the UK. Also, at the train stations, you have many different sounds to give you information. Every time that a train arrives, you hear what is, almost, some kind of theme tune playing. It's like a cartoon theme. Every station has a different sound that gives you some kind of link to the area.

Shoko: Japan

Of all the sounds you can hear, it's the sound of the traffic lights that I can identify as definitely being in Hong Kong. There is a particular sound for the red and green lights. I once experienced a sound installation simulating being blind, where I was asked to listen to the sounds and say where I thought I was. The moment I heard traffic light sounds I was able to identify the location as Hong Kong. I don't think any other city has that same sound. It's almost a sound symbol.

Anonymous: Hong Kong

What are sometimes referred to as 'sensory ethnographic tours' (see Chenhall et al. 2020) have begun to explore soundscape variations within an academic context. These tours, much like public emergency broadcast alerts and rail crossing sounds, are a phenomenon that video-sharing social media platforms are notably fascinated by. A YouTube playlist compiled on the channel *Discord Geographic* provides a set of (at present) 55 short video clips demonstrating pedestrian crossings in China, the Republic of Ireland, UAE, Japan, Taiwan, and Brazil to name a few[5]. These are worth a listen as there are indeed some pretty dramatic differences between these sounds. Armenia, for instance, presents a repeating three-pulse sequence that evokes (in my interpretation) a distinct feel of militaristic marching. Then, with eight seconds remaining before green becomes red, the pattern becomes a constant rapid-fire with a sensation of flange as the tones begin to perceptually merge. The sound is unmistakeably urgent and aggressive, invasive even. A few videos down the list and you can compare this to the sound of a crossing in Japan, in which the green light for pedestrians is accompanied by a two-tone (perfect sixth descending, I checked on my piano) pattern plus a very short, high-pitched beep with a rapidly descending pitch bend that distinctly synthesises the chirping of a bird. The value of aesthetic over urgency is palpable, yet the clarity of the message remains arguably strong.

> Shanghai is definitely noisy. I would describe it as a mixture of many different kinds of sounds. Outside, people's voices are also louder. They don't speak with more emotion or aggression, but even when you have two people standing next to each other, the volume level feels very high, and you're hearing a conversation clearly even if it is meant to be private. Unlike Portsmouth, I certainly wouldn't hear the cries of seagulls and it's unlikely I would hear the sound of a helicopter. In Shanghai, and actually most cities in China, you would rarely hear or see a helicopter. I've also found that in China it is less likely you will hear the sound of pets, specifically dogs, whilst in the UK I think that is a very common sound. When you do hear dogs in China, they are more likely to be louder and sound more energetic and aggressive. I don't know if it's a response to it being busier that makes the dogs more likely to make a lot of noise. I remember, once when my mother visited me in Portsmouth, she commented that the dogs seemed very quiet. She even asked me if British people did something special to their pets to make them quieter.
>
> Wendy: China

Comments regarding communication and vocal expression are explored in the next chapter, but at least one further remark above raised further questions within the interview. This was on the subject of pets, namely dogs, and the observation that they can sound distinctly different between countries. This is particularly difficult to search for in terms of academic and popular literature. The latter interprets seemingly every way of phrasing this as question concerning differences in onomatopoeic language used by humans to describe the sounds of a dog, not the actual sounds produced by the animals themselves. The former has explored the relationship between humans and various pets from an ethnographic perspective (e.g. Anderson 2014; White 2018), but it was difficult to find any reference to sound within such research. If anyone fancies testing the effect of culture upon relative dog-loudness, generating some nice empirical data, I'm quite sure there's a gap.

Soundscapes and routine

Over the last decade or so, research has drawn broad connections between soundscape and daily routine (Schulte-Fortkamp & Genuit 2004). A 2017 study by Craig and colleagues presented evidence suggesting many people have near-identical experiences of sound every day because the soundscape was so deeply connected to their daily routine. The study also revealed that sounds associated with people's daily routines were almost always associated with positive affect and an enjoyable experience.

> When I was a young child, I lived next to a massive factory in Beijing. I don't know what they were making but you could always hear heavy banging and metalwork sounds. I remember, we never needed to use an alarm to keep time, because in the centre of the factory was a huge bell. The factory was open every day apart from Sunday. The bell would first ring at 8am as the workers arrived, then again at midday for the lunch-break, then at 1.30pm as the workers returned from their break. The final bell would ring at 5pm as everyone finished work. A lot of people *not* working in that factory would schedule their day around that bell.
>
> Si: China

Here, the observation concerning factory sounds is notable in that it immediately raised further questions on how soundscapes can influence, if not, benefit the structure of our daily lives.

> Certain sounds have been helping me to start and end my daily activities. Daily, I am woken up at dawn by the sound of The Call to Prayer from the nearest mosque. I listen to Lo-Fi music or brown noise to help me sleep better at night.
>
> Anonymous: Indonesia

> Traffic is a daily sound; the sound of cars when I'm commuting to and from work. After work, what stands out is the sound of notifications on my phone telling me about *Pokémon GO*, which I'm playing at the moment. I also have a notification sound telling me to go to bed.
>
> Anonymous: New Zealand

I have an alarm for the mornings, but I only use it when I have a particular early meeting. Otherwise, I keep it switched off and let myself wake up naturally. The alarm sound itself is definitely one of those sounds that annoys me. It's meant to be a gentle, soothing, musical alarm, but when something wakes you up, it's not soothing.

Jahangir: Bangladesh, UK

From a functional perspective, and reiterating the observations above, aspects of the soundscape can become personalised 'key sounds' that provide structure and, as Brooks and colleagues (2014) observe, can actually "reduce the effort of paying attention" (p.33). I can attest to this personally and have used sound on those cold mornings when even pulling my hands momentarily from my pockets to check my phone or my watch is too much of an imposition. Instead, the quarter-hourly update of the bells atop Portsmouth Cathedral is my personal attenuator in the effort of paying attention, and an architectural and traditional feature of place, borne of religious heritage, is part-structuring my daily routine. Of course, such a relationship with sound is not at all exclusive to a cathedral in Portsmouth or a factory in Beijing. The bells, however, are more of a globally recurrent feature – one that can be indicative of both time *and* space.

> In India we have Hindu temples, which have a lot of bells. During prayer time you have lots of people together, hitting the bells. This produces a distinct sound that you can hear from very far away. The bells sound at specific times of the day and this can give you a sense of your routine. The 5pm bells can signify that it's time to go home for example. In big cities in India, every street has a small temple. At midday they carry out a practice of "giving food to God" where they will hit a bell, and this gives a lot of people the sense that it is now time to go to lunch. The temple bells are also helpful for direction. If you're in India and ask for directions, a lot of the time people will use the temple as a point of reference and then you can use the sound of the bells to help you navigate.

Smita: India

The notion of soundscape for routine also appears below, though on across a wider temporal context of seasonal effects. The nature of the seasons is notably different between countries. Whether a country is land-locked and its proximity to the equator are important factors here. Of course, certain countries, such as Iran, are large enough that they can have dramatically different climates at the same time of the year but between the different regions of the country. Reiterating several of the points raised earlier concerning soundscapes and seasons, the following takes things further by suggesting that certain cultures may actively attend to audible differences to identify and respond to seasonal change.

> Sound in Japan is very seasonal. For example, when we first hear the call of the uguisu, or warbler bird, we know that Spring has come. At first, their call is actually quite short and weak because the bird is not mature.

Their song is not complete yet. As Spring continues, the call becomes longer and more melodic. Then more of the birds start calling to each other and you can hear them communicating. This progression gives you an indication not just that it is Spring, but which stage it is within the season. In the Summer we have the cicada. Their humming becomes a constant background sound. In the earlier part of the Summer, the cicada sounds quite buzzy and aggressive. But as the season continues, their sound softens as their cycle changes and their numbers lessen. As the season moves into Autumn you get a mix of cicadas with crickets, which also sounds very different. These sounds are very meaningful for many Japanese people because the Summers are so hot, the end of the season feels like a relief. The changing sound of the cicada, that marks the end of Summer, is a very positive sound. Understanding these sounds is very helpful to agricultural workers but there is also a general sense of cultural value in having these listening skills.

Shoko: Japan

Most street sellers deal in specific foods. You get used to the tone of each seller's voice in connection to what they are selling. Many sellers also provide seasonal foods, so when they call out their new items this becomes indicative of a change in the season and broadly the time of year.

Smita: India

In Iran, I think that sound can be very different depending on the region you are in and the time of year. For example, there are certain places you can go to hear the sound of waterfalls coming from the mountains. That sound of the water is incredibly relaxing, I think you have to experience it rather than have me explain it. A lot of people believe Iran to be a very hot and dry place, but that's not a fair representation. You have four different seasons. The winter can bring temperatures of minus fifteen or twenty degrees. In the North in particular it's a very mountainous country and you get a lot of snow, whilst in the south it's a lot hotter.

Mehrdad: Iran

Common features, topology, and long-term human influences

Like their European compatriots before them, our contributors' insights into sound and place across Asia and the Middle East have yielded some highly interesting and distinct perspectives and observations. We round off this section with a few final thoughts.

I would describe India as a very noisy country. Particularly if you're living near a road, you can really hear the constant traffic and the big lorries. There are the noises of people selling street food. The sound level is generally high, so people often need to use gesture to help them communicate. In some cases, such as when a person needs to call a rickshaw, they might ask someone else to call out if they feel their own voice won't be loud enough to be heard. When I actually lived there however, I never used to care about the noise. Even if I was studying for an exam and we would have neighbours playing loud music next door, it never affected me. I could study with any regular type of sound around me.

In the big cities such as Mumbai or New Delhi, you experience the sound as a single thing, with every individual sound lost. In a smaller city or town, the lower number of sounds does make things clearer, and you can pick out individual sources of sound much more easily. Around these locations you'll also often hear the sound of the sellers and their trollies. They also operate in the big cities but only in the quieter back streets. Otherwise simply no one would be able to hear them calling. The Indian festivals can be very noisy. In festivals such as Karwa Chauth, which occurs 15 days before Diwali, married women have fasting that ends when you see the moon. A lot of people will set off fireworks at that time, so you know that it is time to end fasting just from the sound.

Smita: India

Jordan is very mountainous. It has a very steep topography with limestone cliffs that effects how sounds are treated. The noise level is always very loud, and the environment is very reflective, so the sound interacts with these surfaces to become even louder. In Amman also, the urbanisation of the city was very rapid, and most of the greenery has been removed. The greenery can buffer a lot of the sound, but pine forests have been totally removed and replaced with buildings, so the sound just reflects everywhere with no obstruction.

In Jordan there are always the sounds of street sellers. They move between the houses, and they will have music playing that will identify what they are selling. Some sellers will speech-sing a repeating phrase whilst others will use a simple instrument melody playing on something similar to a children's toy or music box. There is also a particular celebration that is quite specific to Jordan and some neighbouring countries. It is part of a wedding celebration, but especially celebrates the groom. It happens on the day of the wedding, and all of the local community will come together. The men will walk through the streets together in honour of the groom, and there are specific sticks and drums that the men will carry and hit as they walk. You would immediately recognise this celebration from the sound.

Dana: Jordan

In the capital city of Dhaka, it is very busy and very loud. There's a constant stream of sound from the cars and the horns honking. You can hear a lot of bell jingles from the rickshaws which are everywhere around the city. I don't think there are many sounds that stand out as unique to Dhaka. What feels unique is just the sheer amount of sound as you travel across the city. The etiquette of sound feels different in Dhaka, where there is a greater tolerance for loudness, and you might even say rudeness, in terms of social sound. When people interact outside, I don't get a sense of any heightened emotion, such as anger. People are just louder, which they need to be in order to be heard over the ambient volume of the city.

Jahangir: Bangladesh, UK

A lot of Japan is mountainous. Whilst many people do live in the cities, they are never very far from the mountains and the countryside. Probably 15 minutes by train, so people often visit at the weekends. It's good for children's education to be out in the country and to experience and explore the sound and feel of it. It's also very relaxing for most people. In the winter, the mountains are very quiet. They actually sound sleepy. If you listen very carefully, you can hear the monkeys. Of course, in the spring you get more and more birds, and this makes the sound of the mountains much louder. Insects begin to make sound but they're quieter. Between the spring and the summer, we have the rainy season. The sound of the rain is pretty constant, but gentle, and so you can still hear other things, such as the frogs, that you really notice during this season

Different parts of Japan can have very different natural environments which affects the sound. In Hiroshima, for example, there is almost a sense that the nature is taking over, and you have certain animal sounds that you are unlikely to hear anywhere else. There may be some bears, though their numbers are getting less and less, and you can also hear monkeys and wild boar. In the cities in Japan, it is always noisy. There is a lot of contrast to the countryside. Everything makes noise in the city. I know there is generally a difference between cities and the countryside in countries around the world, but I think that this difference is even stronger in Japan then in, for example, the UK.

Shoko: Japan

Several elements stand out across the above responses, from the unique effects of culturally-specific celebrations and events, to the almost competitive quality of vocalisations out in the world, in which the human voices are not more emotive or aggressive, but greater in loudness as a necessity for being heard due simply to the huge number of voices within the soundscape. Contributors describe a process of acclimatisation to the noise but also to a blurring of the soundscape, with so many sound sources at once the world becomes a cacophony within which individual sound objects and events can become imperceptible. This, in turn, adjusts human behaviour, most notably street sellers who strategically utilise distinctive calls to indicate what they are selling but also are careful to position themselves in quieter locations so that they may increase their chances of being heard and acquiring customers.

Compared to the European leg of our soundscape journey, similar points are raised on observable differences in the boundaries and overlap between urban and rural soundscapes. Topographical effects also feature, notably the reflective effects of cliffs and absorbing impact of the forests and greenery, but also observations on how changes to the built environment are directly impacting upon the natural environment, a process referred to within Pijanowski and colleagues' (2011) soundscape framework as *habitat alteration*. Interestingly, one particular effect not featured in the framework is that our contributor Shoko observes. With reference to Hiroshima, a reversal of habitat alteration is depicted, revealing that the humanisation of soundscapes is not always a permanent or one-way progression. Nature can reclaim a place, a process we could call *habitat reclamation*.

The Americas

Of the studies exploring the soundscapes of the Americas, some focus on environmental impact and the effects various phenomenon are having upon the natural world and the soundscape, including a fair quantity of human activity. Sound is now recognised as an important mediating factor, further affecting wildlife and biodiversity. Research includes Simmons and colleagues' (2021) report on the impact of hurricanes on the coral reef soundscape of the US Florida Keys, and Deichmann and colleagues' (2017) assessment of hydrocarbon exploration on the local soundscapes and the implications to Peruvian bird and frog populations. Other studies look to cultural events and their effects on local people as mediated by sound, such as Medrado and Souza's (2017) damning report *Sonic Oppression*, in which they document the impact the 2016 Olympic Games had upon residents of the *Favela da Maré* in Rio de Janeiro, Brazil. The researchers went so far as to conclude that "[...] militarization is part of a larger government-sponsored program of 'pacification' of the favelas [and] many of the interventions that are being imposed on favela residents manifest in the suppression of sounds" (p.290).

In *Exploring a New Soundscape*, Julie Reid (2019) traces her personal transatlantic soundscape history. For her, San Antonio, Texas was a sonic representation of humid climate, particular flora and fauna, English voices with Spanish tones, and mariachi music that together formed a sonic backdrop that was both languid and festive. For Andrisani (2012), the sound of Havana, Cuba is dynamic, purposeful, and community driven. Here, locals actively contribute to the soundscape with agency, making it "rich in communicative potential, offering the individual a sense of being present within a familiar community" (p.1). Whilst auditory explorations of the Americas appear a little less commonplace than those of Europe and Asia, there is certainly much reason to believe that this particular region of the world is rich and distinct in its soundscapes.

As a state, California is very diverse. There's a wide variety of sounds, from the beaches to the deserts, to the national parks. Many of the sounds here don't really stand out for me because they are my normal. When I travel abroad, I find that sounds stand out more because they aren't my everyday surroundings. When I think of sounds that are dissimilar from place to place, I think of things like varying siren tones and also different sounds from crickets and cicadas. We actually live near a farm, and I can often hear peacocks from the window. It sounds to me like squealing cats.

Michael: USA

I grew up in Mexico City and now live in Malmo, Sweden. And I find that, for example, a park in Mexico does not sound the same in as a park in Sweden. In Mexico, it feels so much more alive with people. You can feel the presence of people through sound. You can feel the life of a community around you. In Mexico it literally sounds hotter. From the sound, you get a sense of a greater blending of social groups which I think links to the culture of Mexico being more open. This social change means that, in Mexico, you hear more sounds from people, so the birds in the trees for example are less noticeable.

Anonymous: Mexico, Sweden

Lima, it's chaos. I think there are two sides to the sound of Lima. On one side, it's a big and busy city with the most condensed population of Peru. There's a lot of informality and there's a sense of disorganisation and competitiveness between the people. On the other side, there is such a mixture of people, and this chaos actually helps create a really rich and unique blend of sound. Peru has several very distinct musical instruments, such as the Peruvian Cajon, which is a percussion instrument that was created and played by the Africans when they were brought over to South America as slaves. Playing the Cajon has given me a sense of relief during difficult parts of my life. When I was young, I would play it every day in my room. My neighbours really hated me for that.

Anonymous: Peru

Competitive loudness that was observed in several countries across Asia makes a reappearance here. Other key features include standout musical tones, resulting from a plethora of uniquely designed musical instruments. City soundscapes imply much of the features identified across Asia, but the emphasis above is very much on the distinct sounds of people. Contributors point to a soundscape that accentuates greater social cohesion and less individualisation of daily life, with consequences for the balance between geophony and anthrophony, as the enhanced social and emotional liberty facilitates more human vocalisation which makes it more difficult to hear the sounds of nature.

Biodiversity and the effects of migration

In addition to the observations above, several responses, presented below, revealed a more consistent highlighting across two elements, both heavily connected to the notion of diversity. The first is the great biodiversity and multiculturality of the Americas. The second is the varied and distinct calls of street vendors and collectors.

The landscape of the Americas is diverse and teeming with exclusive features and characteristics. Across the equator, rainfall envelops the space and the soundscape in equal measure, creating a sonic world many have not experienced even a near-equivalent to. American geography offers several regions in which geophony is almost absolute, with little to no human interference in the soundscape. Those that are present in such space are there to observe – to listen to the sound, not to impact upon it.

> If you go to the coast or other natural areas of Costa Rica, you can observe how biodiverse the country is. We have so many species of birds, insects, and other animals. You can find all kinds of rich sound environments. These soundscapes can be chaotic, but in a symphonic way. Exactly where you are in Costa Rica is going to affect what you hear. Each province is unique in terms of its sound. Even the sound of a single creature, such as the cicadas, changes significantly in tone and persistence between different regions.
>
> The Osa Peninsula has a genuinely rich soundscape because it is mainly a primary forest, completely untouched by man. The sounds from this environment are very different when compared to the tourist locations. The richness of the soundscape really stands out. There is so much sound there but, unlike the city soundscapes, there is a natural harmony and clarity.
>
> There is also some seasonal change in terms of sound, depending on where you are. We have a particular species of paraquet that produces a very distinct sound. They appear at around 4pm in huge groups, making a great ruckus. It's like a symphony of paraquets. But you only hear this for half of the year before they migrate.
>
> Arturo: Costa Rica

> I would describe Venezuela as a very sonic country. We have a very tropical climate. In the summer, around August, we will have a lot of rain, particularly during the evening. I would be with friends at a barbeque for example, and would point out that it's going to rain, not by looking at the sky but by listening to the cicadas. They make a different sound when there is going to be a storm. There is also a lot of diversity between different regions of the country. You can travel along the coast and will be more likely to hear drums. Or you could go towards the Andes mountains where you are more likely to hear Andean influences, such as the Harp, Cuatro and Maracas.

Every country across South America is very different. Here, we are very close to the Caribbean, and so take a lot of influence from there. It's difficult to talk about sound in Venezuela and not talk about music. Music began with the indigenous people, then the colonists came. They brought with them people from Africa as slaves which brought heavy influences from African drum music. The country has also taken immigrants from Europe. Around 1 million Europeans came to Venezuela between 1900 and 1960, mostly from Italy, Portugal, and Spain. There are also communities from the Far East, particularly China, also from The United States, and there is also a big Jewish community here. As these different cultures came into Venezuela, there was little segregation of communities. People mostly integrated over the generations, and they mixed with each other. Historically, when different peoples settle in Venezuela, they do so to escape real difficulties from other places in the world and they settle here permanently. You can hear this in traditional Venezuelan music because it is typically filled with optimism and gratitude. Living in Venezuela in terms of sound means living with a significant range of cultures, customs, accents, and daily activities. It's a really mixed sound.

The Venezuelan capital, Caracas, makes me think of a particular bird call, the Guacharaca. This call is very common in the mornings and has been so for decades. More recently though, there has been a lot of bird immigration from the Amazon because people are trafficking species such as macaws and parrots. Some of these birds were able to escape their confinement, reproduce and spread across Caracas. Now the calls of these birds have started to populate the soundscape of the city.

Andreina: Venezuela

Here, our contributors repeatedly point to a great level of biodiversity, with the chirp and hum of the humble cicada once again making an appearance, but also references to wildlife migration, with characteristically unique species being exchanged between numerous countries across the Americas. As observed in one of our discussions concerning Japan, understanding the local acoustic ecology presents the potential to analyse changes in the soundscape, with tangible benefits to the listener, such as in Andreina's attuning to (once again, and sure to be this book's most featured creature) the cicadas as a means of predicting the weather. The effect seasons have upon the soundscape is also a returning theme, however, here it is explored more so from the perspective of bird migration.

Birdsong of course can describe a wide variety of tones, rhythms, timbres, choruses, and textures; and yet, collectively, denotes what is arguably *the* quintessential sound of nature. Numerous perceptual studies have placed the sounds of birds atop their respective leader boards of listener–preference (Hong & Jeon 2013) and describe our appreciation of it as akin to that music (Rudi 2011). As noted by Jahn and colleagues (2013), more than 230 species of birds born across the southern temperate zone (roughly encompassing the lower halves of Chile and Argentina) migrate north to the tropics to spend the winter months dispersed across various parts of Southern America (Jahn et al. point specifically to the countries of Peru, Brazil, Colombia, and Venezuela). Animal trafficking across South America is less understood when compared to elicit trading across Africa and Asia (Reuter & O'Regan 2017), though, reiterating Andreina's observation, the poaching of wild parrots has been documented in several instances (see Pires 2012 for a full review).

As the commentaries above attest to, the history of migration across the Americas is absolutely not limited to the movement of birds. The discussions placed a clear emphasis on the soundscape effects of multiculturality, with nationalities and cultures colliding on an often-intercontinental scale. Whilst there is resoundingly *not* the scope for even a deeply abridged review of colonialism, multiculturality, and its effects on both soundscapes and auditory perception, there is room to observe that the connection between these elements is absolutely something that has seen study within academic literature. Since 2002, Diwali, the Hindu Festival of Lights, has been an integrated and publicly audible performative event across New Zealand as a celebration of the South Asian communities within the greater population (Johnson 2007). A 2019 study by Tétrault-Farber observes the complex and fluctuating multiculturality of the Montreal soundscape, with Canadian, Québécois, and Irish elements creating what Tétrault-Farber describes as a "vibrant yet precarious milieu" (p.iii). A report on the soundscapes of the humble British Midlands by Rajinder Dudrah (2011) observes how the post-war resettlement of Punjab migrants created a "soundscape of British bhangra" (p.278), as Punjabi folk music became embedded in British culture to the extent that it became a recognised genre of British popular music.

The above examples are without question the very tip of the iceberg. Sound and historical multiculturalism could be (and *is*) the subject of at least one book. The important point to take away, however, is that whilst our relationships with sound can be examined across lines of town, city, country, and continent, humans have been moving around this planet; settling and resettling; dispersing and integrating; and imposing, accommodating, accepting, and opposing each other since the very beginning. Whilst the perceived meaningfulness and scale of these multicultural effects can vary, they are always present.

The calls of commerce

> When I watch some tv shows or movies based in Mexico there are sounds that immediately transport me back, for example the sounds of street vendors, that have a particular recording when they go by, to the music created by 'Organilleros' that are ever present in the city squares, parks and other public places.
>
> Marc: Mexico

Drawing significant overlap with responses from across Asia, the distinct calls from street sellers makes a reappearance in several discussions on the soundscapes of the Americas. Whilst arguably becoming an increasing rarity, though at very different rates across the world, vendors and street sellers using sound to help make themselves standout and reach more customers is an important element of many soundscapes. These sounds may be vocalisations, which themselves may be musical or non-musical in character, or non-vocal expressions using music or other distinctive tones, all with that common goal of attracting custom and making trade. For Garrioch (2003), street sellers were comparable to preachers in their usage of "[the] carrying quality of the human voice [and] appropriate vocal techniques, using pitch, projection and repetition to achieve a high level of audibility" (p.8). Referencing Victor Fournel's, 1858 work, *Ce qu'on voit dans les rues de Paris* (what we see on the streets of Paris), Garrioch gives the example of the 19th-century Parisian coat-seller's call as melancholic yet provocative, whilst the paper-seller across the way would be vigorous and passionate yet clear and classical.

In a review of a video-soundwalk entitled *Sichuan Soundscape* by Emma Zevik, Rees (2002) reviews the soundscape of Sichuan, China, making several observations on the sounds of street sellers. The review notes that the vocalisations of a seller can become associated with that individual, effectively as a brand, particularly if that individual has developed a good reputation from their product or service. In some instances, competing sellers will actually attempt to imitate the vocal call of their successful peer in a deceptive attempt to attain more customers. Rees also observes various non-vocal sounds from street sellers across the Sichuan province, from a food seller throwing their product against metal plates to attract customers by way of the resultant rattle, to the musical instrument seller who, unsurprisingly, uses music – specifically, the piercing yet jovial tones of the fipple flute.

> If you go to the more bohemian parts of Mexico, you can often hear old music being played through a crank handle gramophone. In Mexico City, where I live, we have a market of second-hand buyers and scrap dealers who will collect all kinds of items from people and sell them on. You can find them all through the city and they have a very distinct sound that everyone notices when they visit. It's a recording that wasn't actually produced for this purpose, but one or two sellers started using it, then it caught on and they all started using it. The recording plays from loudspeakers that are usually affixed to the seller's truck. The sound is very loud. It's simply a looping sound of a young girl, I think the daughter of a scrap dealer, saying "we buy", then a series of items, all spoken in Spanish. You will almost always hear this sound throughout Mexico City.
>
> Meni: Mexico

If you'd like to hear that recording described above, it can be dug out online[6]. Whilst becoming less commonplace across the world, vocalisations of street sellers are certainly not gone from the global soundscape. They remain persistent in numerous parts of the world, with no indication that further new technologies or changing attitudes to retail will cause these human expressions to cease. Whilst public perception towards sellers' vocalisations and other sonification approaches can vary, liking for such sounds has generally been evidenced to be high, particularly in urban areas that are regarded as 'historical' (Zhou et al. 2014) – potentially due to greater levels of expectation and perceived appropriateness in such districts (Bahali & Tamer-Bayazit 2017). In many cases, the contribution to the soundscape provided by street sellers is something that has been identified as an important auditory heritage that must be preserved (Brambilla & Maffei 2011).

Africa

In Alexandria, Egypt, something there that immediately comes to mind is the sonic vibrance of the place. It's something more than simply the sound of a big city. In Alexandria, everything is noisy, but it is noisy because it *needs to be*. The people need to be noisy so that they can be seen, heard, and perceived. The traffic is a constant flux of horns and voices, but actually in a rhythmic way. It's overwhelming but somehow, it's also very comforting, like an immersive blanket of sound. Of course, being in Egypt, you can drive a few miles out of the city, towards the desert. Here you can find absolute silence, and, in that moment, the absence of sound is a blessing before you return to the cacophony of the city.

Ludo: Italy, Egypt, UK

Although attention to the soundscapes of Africa is a relatively new endeavour, at the very least when compared to European and Asian explorations, more recently published soundscape research suggests that this is steadily changing. For de Witte (2016), the city of Accra in Ghana, "is an intense sonic experience [...] alive with sounds [and] cacophonies of voices: talking, calling, shouting, hissing, bargaining, quarrelling, laughing, singing, preaching" (p.133). As observed by Sinamai (2018), the powerful folklore and cultural heritage of Great Zimbabwe coalesces into a soundscape that is steeped in sacred meaning and auditory memories so vivid, they are more of an act of re-experiencing than remembering. Further soundscape research examining specific countries and locales within Africa remains difficult to find, with the more notable studies exploring matters of sound and perception across the continent more broadly (e.g. Kankhuni & Ngwira 2021). Therefore, there is much room here for further soundscape exploration.

There is a lot of different sound in terms of nature across Kenya. With these sounds, you know that you are in Africa. Just as if I wake up in a village in England I can tell just from the sound where I am, it's the same in Kenya. For example, the sound of a cockerel crowing is the defining sound of waking up in Kenya. I hear that sound and I know I'm in Kenya. I know there are certainly chickens and cockerels in England but I don't draw the same connection between that sound and the UK.

Anonymous: Kenya

As we reach the final leg of our continent-hopping review of how the world sounds, let us close with some of the remarks made by our fine contributors upon the continent of Africa, that emphasised what became two emergent themes: tradition and spirituality, and the sounds of transport.

Tradition and spirituality

In Togo, when the traditional chiefs have information to communicate, they use sound. In some villages, not everybody has access to a television and so the chief's messengers go to every corner of the village beating on a percussive musical instrument called 'gankogui' or 'agogo'. When the messengers hit the gankogui in a certain way, this tells everyone that there is an important message to announce, and everyone should gather around to listen.

In both Togo and Senegal, we don't have clear seasonal changes like winter or summer, but there are certain points during the year where the cultural environment can change, and this can affect the sound. During the one month fasting period of Ramadan for example, the streets become a lot quieter and there is less noise from the people. You actually notice more sounds of nature during this time. The birds and the wind both stand out more. In Senegal, you would expect to hear the sound of the tam tams to identify a significant event. This sound has been used to notify the community when someone has died, but also events such as baptisms. Recently, the tam tams have been used more for positive events with people choosing not to use the sound in negative contexts.

There is also a connection between sound and religion in Senegal. Here, there are many Islamic schools where children learn the Qur'an mainly by singing it. For some people, they say that they can feel more, and they experience the strength of the words when they are sung rather than spoken, and this helps the children to retain the knowledge from their readings. In Senegal there is also a form of spiritual singing called "khassaïde" that will bring people together. This can happen almost anywhere, not just at a designated building. It could be outside on the street and could happen at any time, though it is more common in the early evening.

Anonymous: Senegal, Togo

Across the African soundscape, religious influence makes a reappearance but here it is connected to almost season-like effects, with certain practices taking place over particular, extended periods of time. As described, this affects the balance of geophonic and anthrophonic sounds within the soundscape during that period, with reductions in the human-created sounds presenting space within the soundscape for the sounds of flora and fauna to emerge. Further examples are given of unique musical instruments that present distinct musical timbres, textures, and tonalities. The notion of 'competitive loudness' is also a recurrent theme. Of course, there are also numerous features distinct from that which we have already uncovered. Wonderful examples include the spiritual singing of Khassaïde[7], the use of sound by village chiefs to herald announcements and gather people together to deliver important messages, and the use of singing specifically as a means of supporting memory and retaining information. The latter of these, in particular, has notable backing within scientific literature, with various studies attesting to the power of singing to facilitate learning, particularly in contexts such as early years of development (Ginting 2019) and when learning a foreign language (Good et al. 2015).

Transportation

The public transport system in Nairobi has a lot of van-taxis. When you are out on the street it really is loud. London buses are so much quieter. I've started cycling to work and I honestly cannot hear them coming sometimes, whilst in Africa, I'd expect to hear a bus coming from a mile away. Also, the way the mini-van taxis call to customers in Nairobi is through music. They play music loudly as there's this sense of the more loudly you play your music, the better sound system you have and therefore the more customers you will attract. So there ends up being this competition between taxis over who can play their music the loudest. This is the same for the touts in the city markets who are often shouting their words and making jokes. People are again, more likely to go into the stalls and shops that are louder. There's a saying, 'you know you've become old when you decide to pick a quieter taxi'. These taxis are called matatus and some are quieter, but if you pick one of these you are seen socially as 'old'.

Nairobi sounds alive. There's no doubt that there are people here whilst, in London, I feel like the cars are driving themselves. There's traffic in London, but to me it almost feels silent. The sound in Nairobi is certainly chaotic, but it's alive with people. You can tell how people are feeling through the sound. The way people drive their cars, the sounds of the engine and they are using their horns more. In Kenya, as well as emergency services, several dignitaries and important people are also allowed to have sirens on their vehicles and use these to move traffic out of their way. Kenya has moved from a more centralised system to one that is county-based. This has meant that whilst previously maybe only the President and Vice-President would have sirens on their vehicles, now there are a lot more of these important people who feel they need sirens also.

Anonymous: Kenya

How sounds of the city could be dictated by transport was an intriguing research proposition. The presence of an underground metro system within a city has been shown in some instances to be a negative influence upon the perceived quality of an urban soundscape. As Yilmazer and Bora (2017) observe in their assessment of the Akköprü metro station in Turkey, respondents were notably more likely to describe the interior soundscape of the metro as unpleasant and stressful whilst adjacent outdoor space was described using considerably more positive language. The particular qualities of public transport will of course be heavily linked to multiple factors, not least infrastructure, but also numerous cultural and social elements. Returning to de Witte's (2016) soundscape review of Accra, so-called "roaming evangelists" (p.133) are commonplace on public buses and will use dramatic vocalisations or even full-volume loudspeakers to present their captive audience with the word of God. This isn't something you're particularly likely to hear on the Number 18 bus service between Fratton Park and South Parade Pier.

A brief epilogue: sound everywhere and nowhere

British born and raised in the UK, one particular interviewee confounded the notion of place in their interview. An officer in the Royal Navy, their commentary largely centred upon life with sound onboard a ship. This ship could be almost anywhere, at any point in time. Despite this, the chance to gain some insight into sound and place, when the place in question was constantly moving across the world, was too good an opportunity to miss.

> On the ship, there is always the constant sound of air conditioning in the cabin. When we're at sea, you will hear the constant rumbling of the diesel engines and the creaking of the ship's structure as it moves. There is a constant, low-level of ambient noise throughout the ship but it's not uncomfortable. I wouldn't say there is much variation in sound between ships. The announcements on the ship's public address speakers are not constant but certainly a very regular source of sound, as are telephones ringing and the whirring fans of all the computers. If I concentrate, I can pick up the more subtle constant hums of the light bulbs and the sounds of television through the walls in an adjacent room. In dock, there's also the persistent sound of traffic that we can hear even when inside the ship. There's often a train station close to the harbour so you will frequently hear the sound of trains arriving and departing. In Portsmouth this stands out as, here in particular, they are bloody squeaky. The sound is so obtrusive, it got to the point where I found myself noticing that, although both incoming and departing trains squeak, those entering the station were reliably squeakier.
>
> There are certain areas of the ship, such as the engine room, where the volume levels are so high that you either need to avoid them or wear protective equipment. I also think human interaction and communication is

a big part of sound experience on board. Unless I make it clear that I am not to be disturbed, there will be very regular knocks at my cabin door. My phone will constantly be ringing. Such sound is definitely the norm, and if you don't want that then you have to take action to avoid it. One perspective that is certainly anecdotal but also very popular, is that because we operate in a very loud ambient sound environment, that is what we become used to. If I go home to my house, it almost feels unnatural in terms of sound, because there is so much less ambient noise. Colleagues have often told me that they struggle with sleeping at home, precisely because it is too quiet and there isn't that same level of ambient noise that your body has become accustomed to.

One particularly common sound is that of someone using the public address system incorrectly. They will hold the microphone too close to their mouth, or too close to a nearby speaker and you'll get feedback – this screeching sound distorting the messages. There is also a basic coding system we use for the ship's horn, where a certain combination of toots means specific things to other ships that other drivers know how to interpret. Of course, this becomes very important in foggy conditions.

Anonymous: UK

In 1912, a retrospective on the life's work of the renowned cinematographer, and Portsmouth native, Alfred West was published. Entitled *Life in our Navy and Our Army*, the work prioritised the visual medium but was most certainly not dismissive of the role of sound within this context. The text references numerous sound objects and events indelibly tied to naval life. The deep chimes of Eight Bells Noon that sets off hunger pangs across the ship with its announcement of the midday meal. The shrill blare of the whistle as the evocative sonification of 'go!'. The quick-march rhythm of the bugle denoting an imminent drill. The bugle features in many instances, sounding for all manner of communique, from "Man and Arm Ship" signalling every sailor to take up position behind their gun, to "All hands on the deck house", sounded in the event of an enemy vessel ramming into the ship.

A 2006 report for the US Navy (Bowes et al.) revealed that for the previous financial year, $137 million had been paid in compensation to naval veterans for their suffering of hearing loss. Though this paled in comparison to the $475 million cost to the US Army, the report prompted the Navy to reconsider its perspectives on sound and investigate means of removing or attenuating sources of loud noise. Specific recommendations included the creation of quiet spaces on board ships for recovery, increased availability of personal noise-protection devices (such as earplugs and overhead earmuffs), and more comprehensive integration of hearing protection as part of health and safety training. Whilst this does consider sound and health from a physiological perspective, more psychological matters and affective wellbeing appear to have gathered less attention within literature, but they are shown some appreciation in practice.

On board the ship, people do shift work and so you are expected to be respectful of people who have come off a night shift for example. Staff working the 12–4am shift can sleep until 10am, whilst the majority of the crew will start work at 8am. During this overlap, people are expected to be considerate towards those who are still asleep so you might expect there to be a minimising of public address announcements during this time. The same etiquette applies after 10.30pm, again because people are expected to be sleeping.

Anonymous: UK

Outside of the above and a handful of other examples, research addressing sound within a military context, specifically non-musical sound, is notably limited. As described earlier in this book, one important application of soundscape research is improvement of wellbeing and quality of everyday life, and this lack of research in a military context presents a gap in our understanding and an opportunity to build a greater awareness of the professional and personal relationships between military personnel and sound, to ultimately better-support their wellbeing whilst they are operating in some of the modern world's most challenging and dangerous circumstances.

Chapter 5 summary: there's a lot going on

Using place as a means of structuring an exploration of soundscape is nothing if not problematic. Soundscapes across the world present us with a hugely diverse range of features and qualities that invite observations of both similarity and difference at probably every level of scale. The degree to which a particular place can be described as homogenous and generic or heterogenous and unique is both subjective and tough to benchmark. This remains true whether you are attempting to compare continents, countries, regions, cities, countryside, commercial and domestic locales, or tourist attractions. That said, there is most certainly great value in obtaining perspectives on soundscape from a collection of contributors who hail from all four corners. Crucially for the interviews, the various topics that emerged from this initial question of sound and place prompted numerous follow-up questions as we explored further important issues of the human relationship with sound. Matters of routine led to discussions upon how sound features in the structure of everyday life, whilst commentary on human vocalisation led to much conversation on matters of 'social sound', etiquette, and consideration of one's own sound within a social context. We explore these matters in the next chapter.

Notes

1 Direct link to Herranz-Pascual and colleagues' (2010) conceptual framework for studying soundscapes: https://www.researchgate.net/profile/Igone-Garcia/publication/285200832 (accessed 07.02.2022).

2 Illustration of Pijanowski et al.'s (2011) soundscape ecology framework: https://media.springernature.com/lw685/springer-static/image/art%3A10.1007%2Fs10980-011-9600-8/MediaObjects/10980_2011_9600_Fig2_HTML.gif (accessed 19.02.2020).

3 News reports covering the mark left behind by supporters in the wake of England losing the 2020 European Cup final: [https://metro.co.uk/2021/06/19/clean-up-begins-after-football-fans-trash-leicester-square-14798148/] [https://www.itv.com/news/2021-07-12/big-clean-up-after-england-football-fans-head-home-from-euro-2020-final] (accessed 05.02.2022).

4 For anyone who is interested and hasn't heard a simulated engine before, YouTube has a bit of a fascination with this topic, particularly in high-performance cars, https://www.youtube.com/watch?v=xX5Phiu623s for example (accessed 06.02.2022).

5 Discord Geographic (YouTube channel) compilation of traffic/pedestrian crossing lights and sounds: https://www.youtube.com/playlist?list=PLYNzgrWx34qqouUbMcMcfh6_cnWaeg9L5 (accessed 08.02.2022).

6 Link to the Mexico City scrap metal collector sound: https://www.youtube.com/watch?v=x3zNGTVGv4s (accessed 19.02.2022).

7 Example of Khassaïde spiritual singing: https://www.youtube.com/watch?v=M-ljtR2l41_k (accessed 08.02.2022).

References

Adams, M., Cox, T., Moore, G., Croxford, B., Refaee, M., & Sharples, S. (2006). Sustainable Soundscapes: Noise Policy and the Urban Experience. *Urban Studies*, 43(13), 2385–2398.

Aletta, F., Kang, J., & Axelsson, Ö. (2016). Soundscape Descriptors and a Conceptual Framework for Developing Predictive Soundscape Models. *Landscape and Urban Planning*, 149, 65–74.

Anderson, P. K. (2014). Social Dimensions of the Human–Avian Bond: Parrots and Their Persons. *Anthrozoös*, 27(3), 371–387.

Andrisani, V. (2012). Aural ethnography and the notion of membership: an exploration of listening culture in Havana. Latin American Studies Working Paper Series. Simon Fraser University. http://summit.sfu.ca/item/10795 (accessed 09.02.2022).

Bahalı, S., & Tamer-Bayazıt, N. (2017). Soundscape Research on the Gezi Park–Tunel Square Route. *Applied Acoustics*, 116, 260–270.

Benderius, O., Berger, C., & Lundgren, V. M. (2017). The Best Rated Human–Machine Interface Design for Autonomous Vehicles in the 2016 Grand Cooperative Driving Challenge. *IEEE Transactions on Intelligent Transportation Systems*, 19(4), 1302–1307.

Bernat, S. (2014). Soundscapes and Tourism–Towards Sustainable Tourism. *Problemy Ekorozwoju–Problems of Sustainable Development*, 9(1), 107–117.

Bowes, M. D., Shaw, G. B., Trost, R. P., & Ye, M. (2006). *Computing the Return on Noise Reduction Investments in Navy Ships: A Life-Cycle Cost Approach*. Center for Naval Analysis (CAN) Report D, 14732.

Brambilla, G., & Maffei, L. (2011). Soundscape Heritage: An Evolving Value to Preserve and Archive? Soundscape as a Part of Cultural Heritage [COST Action TD0804].

Chenhall, R., Kohn, T., & Stevens, C. S. (2020). *Sounding Out Japan: A Sensory Ethnographic Tour*. Abingdon: Routledge.

Christoforou, Z., de Bortoli, A., Gioldasis, C., & Seidowsky, R. (2021). Who Is Using e-Scooters and How? Evidence from PARIS. *Transportation Research Part D: Transport and Environment*, 92, 102708.

Clark, T. (2006). I'm Scunthorpe 'Til I Die': Constructing and (Re) Negotiating Identity Through the Terrace Chant. *Soccer & Society*, 7(4), 494–507.

Colombijn, F. (2007). Toooot! Vroooom! The Urban Soundscape in Indonesia. *Sojourn: Journal of Social Issues in Southeast Asia*, 22(2), 255–272.

Craig, A., Moore, D., & Knox, D. (2017). Experience Sampling: Assessing Urban Soundscapes Using In-Situ Participatory Methods. *Applied Acoustics*, 117, 227–235.

de Witte, M. (2016). Encountering Religion Through Accra's Urban Soundscape. In: Darling, J. & Wilson, H. (Eds.) *Encountering the City* (pp. 133–150). Abingdon: Routledge.

Deichmann, J. L., Hernández-Serna, A., Campos-Cerqueira, M., & Aide, T. M. (2017). Soundscape Analysis and Acoustic Monitoring Document Impacts of Natural Gas Exploration on Biodiversity in a Tropical Forest. *Ecological Indicators*, 74, 39–48.

Duarte, C. M., Chapuis, L., Collin, S. P., Costa, D. P., … & Juanes, F. (2021). The Soundscape of the Anthropocene Ocean. *Science*, 371(6529), eaba4658.

Dudrah, R. (2011). British Bhangra Music as Soundscapes of the Midlands. *Midland History*, 36(2), 278–291.

European Environment Agency. (2014). *Good Practice Guide on Quiet Areas*. Publications Office of the European Union, Bonnevoie, Luxembourg.

European Parliament and Council. (2002). *Directive 2002/49/EC Relating to the Assessment and Management of Environmental Noise*. Brussels: Publications Office of the European Union.

Faas, S. M., & Baumann, M. (2021). Pedestrian Assessment: Is Displaying Automated Driving Mode in Self-driving Vehicles as Relevant as Emitting an Engine Sound in Electric Vehicles?. *Applied Ergonomics*, 94, 103425.

Fournel, V. (1858). *Ce qu'on voit dans les rues de Paris*. A. Delahays.

Garrioch, D. (2003). Sounds of the City: The Soundscape of Early Modern European Towns. *Urban History*, 30(1), 5–25.

Gebhardt, L., Wolf, C., & Seiffert, R. (2021). "I'll Take the E-Scooter Instead of My Car"—The Potential of E-Scooters as a Substitute for Car Trips in Germany. *Sustainability*, 13(13), 7361.

Ginting, M. B. (2019). Improving the Memory Through Singing Method of Children Ages 5–6 Years in Kindergarten Insan Pandhega. *International Journal of Emerging Issues in Early Childhood Education*, 1(2), 94–107.

Good, A. J., Russo, F. A., & Sullivan, J. (2015). The Efficacy of Singing in Foreign-Language Learning. *Psychology of Music*, 43(5), 627–640.

Gössling, S. (2020). Integrating e-Scooters in Urban Transportation: Problems, Policies, and the Prospect of System Change. *Transportation Research Part D: Transport and Environment*, 79, 102230.

Halliday, W. D., Pine, M. K., Mouy, X., Kortsalo, P., Hilliard, R. C., & Insley, S. J. (2020). The Coastal Arctic Marine Soundscape Near Ulukhaktok, Northwest Territories, Canada. *Polar Biology*, 43(6), 623–636.

Hayne, M. J., Rumble, R. H., & Mee, D. J. (2006). Prediction of Crowd Noise. In: McMinn, T. (Ed.) *Proceedings of the First Australasian Acoustical Societies Conference* (pp. 176–188). Christchurch: New Zealand Acoustical Society.

Herranz-Pascual, K., Aspuru, I., & García, I. (2010). *Proposed Conceptual Model of Environmental Experience as Framework to Study the Soundscape.* Lisbon: InterNoise.

Hong, J. Y., & Jeon, J. Y. (2013). Designing Sound and Visual Components for Enhancement of Urban Soundscapes. *The Journal of the Acoustical Society of America*, 134(3), 2026–2036.

Jahn, A. E., Levey, D. J., Cueto, V. R., Ledezma, J. P., Tuero, D. T., Fox, J. W., & Masson, D. (2013). Long-Distance Bird Migration Within South America Revealed by Light-Level Geolocators. *The Auk*, 130(2), 223–229.

Johnson, H. (2007). 'Happy Diwali!' Performance, Multicultural Soundscapes and Intervention in Aotearoa/New Zealand. *Ethnomusicology Forum*, 16(1), 71–94.

Kang, J., Aletta, F., Gjestland, T. T., Brown, L. A., Botteldooren, D., Schulte-Fortkamp, B.,… & Lavia, L. (2016). Ten Questions on the Soundscapes of the Built Environment. *Building and Environment*, 108, 284–294.

Kankhuni, Z., & Ngwira, C. (2021). Overland Tourists' Natural Soundscape Perceptions: Influences on Experience, Satisfaction, and Electronic Word-Of-Mouth. *Tourism Recreation Research*, 46, 1–17.

Kelman, A. Y. (2010). Rethinking the Soundscape: A Critical Genealogy of a Key Term in Sound Studies. *The Senses and Society*, 5(2), 212–234.

Konet, H., Sato, M., Schiller, T., Christensen, A., Tabata, T., & Kanuma, T. (2011). Development of Approaching Vehicle Sound for Pedestrians (VSP) for Quiet Electric Vehicles. *SAE International Journal of Engines*, 4(1), 1217–1224.

Kytö, M. (2011). 'We Are the Rebellious Voice of the Terraces, We Are Çarşı': Constructing a Football Supporter Group Through Sound. *Soccer & Society*, 12(1), 77–93.

Liu, A., Wang, X. L., Liu, F., Yao, C., & Deng, Z. (2018). Soundscape and Its Influence on Tourist Satisfaction. *The Service Industries Journal*, 38(3–4), 164–181.

Maiti, A., Vinayaga-Sureshkanth, N., Jadliwala, M., Wijewickrama, R., & Griffin, G. P. (2019). Impact of E-Scooters on Pedestrian Safety: A Field Study Using Pedestrian Crowd-Sensing. In: Passarella, A. & Vallati, C. (eds.) *2022 IEEE International Conference on Pervasive Computing and Communications Workshops and other Affiliated Events* (pp. 799–805). New York: IEEE.

Marra, P. S., & Trotta, F. (2019). Sound, Music and Magic in Football Stadiums. *Popular Music*, 38(1), 73–89.

Medrado, A., & Souza, R. (2017). Sonic Oppression, Echoes of Resistance and the Changing Soundscapes of Rio's Favelas in the Build-Up to the Olympics. *Journal of Radio & Audio Media*, 24(2), 289–301.

Negrín, F., Hernández-Fernaud, E., Hess, S., & Hernández, B. (2017). Discrimination of Urban Spaces with Different Level of Restorativeness Based on the Original and on a Shorter Version of Hartig et al.'s Perceived Restorativeness Scale. *Frontiers in Psychology*, 8, 1735.

Philpott, C. (2013). The Sounds of Silence: Music in the Heroic Age of Antarctic Exploration. *The Polar Journal*, 3(2), 447–465.

Philpott, C., & Leane, E. (2021). The Silent Continent? Textual Responses to the Soundscapes of Antarctica. *ISLE: Interdisciplinary Studies in Literature and Environment.* 28(1), 15–29.

Pijanowski, B. C., Farina, A., Gage, S. H., Dumyahn, S. L., & Krause, B. L. (2011). What Is Soundscape Ecology? An Introduction and Overview of an Emerging New Science. *Landscape Ecology*, 26(9), 1213–1232.

Pires, S. F. (2012). The Illegal Parrot Trade: A Literature Review. *Global Crime*, 13(3), 176–190.

Polli, A. (2012). Soundscape, Sonification, and Sound Activism. *AI & Society*, 27(2), 257–268.

Rees, H. (2002). Sichuan Soundscape. *Asian Music*, 34(1), 167–170.

Reid, J. (2019). Exploring a New Soundscape. Explore Life.com. Online article: https://www.explore-life.com/en/articles/exploring-a-new-soundscape (accessed 06.02.2022).

Reuter, P., & O'Regan, D. (2017). Smuggling Wildlife in the Americas: Scale, Methods, and Links to Other Organised Crimes. *Global Crime*, 18(2), 77–99.

Rudi, J. (2011). Soundscape and Listening. *Soundscape in the Arts*, 12(2), 185–194.

Schafer, R. M. (1970). *The Book of Noise*. Wellington: Price Milburn.

Schafer, R. M. (Ed.). (1977). European Sound Diary (No. 3). ARC Publications: ARC The Aesthetic Research Centre: Burnaby: World Soundscape Project.

Schulte-Fortkamp, B., & Genuit, K. (2004). The Acoustical Diary as an Innovative Tool in Soundscape Evaluation. *Journal of the Acoustical Society of America*, 115(5), 2496.

Schulte-Fortkamp, B., & Jordan, P. (2016). When Soundscape Meets Architecture. *Noise Mapping*, 3(1), 216–231.

Sherriff, G., Blazejewski, L., Hayes, S. J., Larrington-Spencer, H. M., & Lawler, C. (2021). E-Scooters in Salford: Interim Report.

Simmons, K. R., Eggleston, D. B., & Bohnenstiehl, D. R. (2021). Hurricane Impacts on a Coral Reef Soundscape. *Plos One*, 16(2), e0244599.

Soares, F., Silva, E., Pereira, F., Silva, C., Sousa, E., & Freitas, E. (2021). To Cross or Not to Cross: Impact of Visual and Auditory Cues on Pedestrians' Crossing Decision-Making. *Transportation Research Part F: Traffic Psychology and Behaviour*, 82, 202–220.

Tańczuk, R., & Wieczorek, S. (2018). *Sounds of War and Peace: Soundscapes of European Cities in 1945* (p. 270). New York: Peter Lang International Academic Publishers.

Tétrault-Farber, J. (2019). Une ville-plusieurs reels: Montreal's Multicultural Irish soundscape (Doctoral dissertation, Concordia University).

To, W. M., Chung, A., Vong, I., & Ip, A. (2018). Opportunities for Soundscape Appraisal in Asia. In Taroudakis, M. (Ed.) *Proceedings of the Euronoise 2018 Conference*, Euronoise, Crete, Greece.

Wells, H. L., McClure, L. A., Porter, B. E., & Schwebel, D. C. (2018). Distracted Pedestrian Behavior on Two Urban College Campuses. *Journal of Community Health*, 43(1), 96–102.

West, A. (1912). *Life in Our Navy and Our Army*. Portsmouth: Wessex Press.

White, P. (2018). (Un) Regulated Relations: An Ethnographic Perspective of Dog Ownership on Isla Mujeres, Mexico. *Anthrozoös*, 31(5), 615–625.

Yilmazer, S., & Bora, Z. (2017). Understanding the Indoor Soundscape in Public Transport Spaces: A Case Study in Akköprü Metro Station, Ankara. *Building Acoustics*, 24(4), 325–339.

Zhou, Z., Kang, J., & Jin, H. (2014). Factors That Influence Soundscapes in Historical Areas. *Noise Control Engineering Journal*, 62(2), 60–68.

6 Sound and everyday experience

How sound functions has been explored at various points in previous chapters, both in terms of academic research and contributor comments. Here we take a more focussed look at the nature and function of sound within the 'everyday'. First, an exploration of sound within daily routine. Then, our second port of call considers various questions on how sound features within the home. This is followed with a return to R.M. Schafer's most subjective of inquiries into the beauty of sound, as we discuss our contributor's thoughts on 'good' and 'bad' sound within their everyday experience. We conclude with a discussion on the interpersonal aspects of everyday sound, as we ask what is 'sound etiquette'?

DOI: 10.4324/9781003178705-7

How does sound function in your everyday life?

Background sound

When asked about the role of sound in everyday life, their daily routine, in particular, many respondents pointed to sound as 'something of the background'; a near-constant companion that they can attentionally drift in and out of whilst attending consciously to other things as they go about their day.

> When I'm cooking, I like to have the tv or radio on. This could be music or just the news or another programme. Anything is fine, but I like to have that background of sound when cooking.
>
> Wendy: China

> I go to the gym in the mornings, where I will listen to some extended aggressive music. Back at home I listen to podcasts before I start work. After work I listen to podcasts again, or more relaxed music, while I do stuff around the house. There are almost no moments without a background of sound.
>
> Anonymous: Germany

> I sometimes use certain television shows to help me go to sleep. I use the automated, 30-minute shut off so I can fall asleep in the first 10 minutes, then the system turns itself off. Sound that I'm familiar with can help me to fall to sleep. Something that can calm me down, to distract from the conscious, critical thinking around my work. I'm constantly thinking about emails and other parts of my job. If you don't have any sound to take away from this thinking, that can be very disturbing for sleep. Just shutting your eyes doesn't shut out these thoughts. The sound from the television, these are programmes I've seen many times before. I don't need to concentrate on them or feel involved in the fictional situation. The sounds of the voices especially helps to send me to sleep. It's effectively a medicine.
>
> Si: China, UK

> If I can't sleep, I tend to use the sound of television programmes. It has to be a particular type of programme and it's the voices that help me to sleep. If there's theme music, I have to skip that as it's often too upbeat and makes me more awake. Often, it's comedy programmes, but ones that I've seen so many times that I don't actually have to concentrate on what's being said. I'm half-listening. Because I already know what's going to be said, I actually find it really comforting. You don't even need to watch, you just use the sound and the voices, and you know what's happening. It's really calming. Even when I'm not trying to sleep, I still find it comforting to experience familiar television programmes in this

way. This same sound of television as background noise I also find very important whilst I'm eating. I absolutely need to have the programme I'm going to watch all set and ready from the moment I sit down to eat. I think that comes from a habit I picked up as a student living in halls of residence. Again, it's familiar sound that's comforting. It reinforces the idea of it being my time.

Laksha: Sri Lanka, UK

When I need to concentrate, listening to recordings of natural sounds can help. These might be rain sounds or thunderstorms. White noise-type sounds that help cancel out other noises and distractions. When I need to feel more energetic, or positive if I feel sad, then I'll often use specific music to regulate my emotional state. I also find that if I want to relax at home for example, then trying to create a silence is very effective.

Lucia: Spain

I often fill my home with sound, whether it's the radio or the television, I will typically have something producing sound or music throughout the day. I do enjoy silence in some respect, but I think it's a bit like chocolate in that I'm not in the mood for it all day or every day. It is rare that the house will be silent. There's always some kind of buzz going on.

Anonymous: Italy

Using specifically selected sound as background ambience is of course a very common feature, particularly in the home, both first thing in the morning whilst preparing for work, and later in the evening as part of a routine of relaxation. Music is another commonality. However, most respondents did not go into great detail regarding the specifics of their music choices. Conversely several indicated quite the opposite, noting that the exact music doesn't particularly matter, it's simply that there *is* music accompanying them in their routine that is important. An appreciation of silence also features prominently within certain contexts, most notably when trying to fall asleep – though this raises a clear point of division between many responses.

One particular observation that I personally found to be highly relatable was the television being used exclusively for its sound. Whether as a means of falling asleep, accompanying dining or the preparation of a meal, or supplementing other work or recreational activity, many contributors pointed to the soundtrack of the television as a way of augmenting their homes with familiar, reassuring, and predictable sound. The final comment above also, tellingly, uses the phrase "fill my home with sound", emphasising the three-dimensional, resonant qualities of sound which, unlike the visual image, is not fixed to a screen. It expands outwards, interacting acoustically with the space and the objects within. There is an immersive, enveloping quality to sound that could explain part of our attraction to this usage of 'television as background noise'.

Situational awareness

Matters of how sound supports situational awareness are explored as part of the discussion on sound affordance within Chapter 3. However, many of our contributors raised some interesting points that overlapped situational awareness with everyday experience, observing several distinctive ways in which sound communicated information to them.

> In Malaysia, but also countries such as Singapore, Vietnam, and Thailand, we have what's called a 'wet kitchen'. Asian cooking can involve a lot of deep frying and steaming where you're going to get a lot of condensation and grease on walls and surfaces. This is done in a wet kitchen, whilst a dry kitchen is more for heating already cooked food or preparing lighter dishes. The wet kitchen is almost always right at the back of the house where it can be ventilated with big windows, which of course also lets out all the noisy sounds of your cooking to the neighbours. If I hear the sound of flames or the wok being used in my neighbour's kitchen, I immediately know that something delicious is being prepared. I also find that by combining the sound of the activity in the kitchen with the smells, I can usually identify what kind of meal is being prepared, and this can influence what I will have when I make my next meal. There's definitely a behavioural reaction to the sound that I hear.
>
> Anonymous: Malaysia

> When I am cooking, I don't think about what grade the extractor fan is on, I can just tell by listening. I think that sound in general helps me to cook, and probably helps me to not burn my house down. My intercom – I use that a lot. The problem is that the buzzer that sounds downstairs when I unlock the main door is quite far away from the intercom, so a lot of the delivery drivers new to the building don't realise that I've opened the door and they just keep buzzing. So, the exact pattern of sound from the person buzzing actually tells how experienced the delivery driver is.
>
> Anonymous: Kenya

> I can recognise individual members of my family around the house by the sounds of their movement. The sound from the shoes they are wearing tells me who is there. My mother, for example, wears an anklet with a series of beads that produce a very specific sound when she walks. There's also a certain rhythm in different people's steps. I also notice that you can identify a person by the sound of their routine or activity. When someone uses the bathroom, the order or duration with which they make certain sounds can tell you who the person is. There's even difference in sound in the same action, depending on the person. When my neighbours open their windows, the husband opens the window loud and fast. When his wife does the same, the sound is lighter and smoother.
>
> Smita: India

I sometimes consciously use sound when I'm driving. I find that having music on in the car can actually make me feel safer because it helps put me at ease and this helps me to focus. When driving, I can get a bit overwhelmed and so a more familiar and constant sound from the music is calming. There are a lot of driving sounds, such as the reverse warning beep alarm, that I absolutely hate. These buzzers and indicators always make me feel there's something wrong. Of course, sound is really helpful when you have limited visibility. Reversing out onto a road where there's a visual obstruction for example. I will always wind down my window to listen out for oncoming vehicles. I also insist that my partner roll down their side window too, so that I can listen for traffic approaching from both sides.

There's a lot of sound that features in cooking. When my mum taught me how to cook the base of a curry, for example, she showed me how to listen out for the pop of the mustard seeds before you put the fennel seeds in. I use a pestle and mortar a lot, and that sound changes as you grind so you can again, use your ears to help direct your cooking process. There are various ways in which sound can indicate how well things are going in the kitchen. The particular sound of the pot bubbling at just the right amount, or the sound of a little test-batter in oil to check you're cooking at the correct temperature, it all helps with technique and timing in your cooking.

Laksha: Sri Lanka, UK

I tend to avoid headphones. I don't like to have my ears covered. I don't think that's always safe. I want to be able to hear if my doorbell rings or to know what's happening whilst I'm walking. A lot of people are happy to wear headphones when walking home or to work, but if a bicycle was passing by and I needed to give way, I wouldn't be able to hear them if I was wearing headphones. If I can hear the sounds around me, I am more able to react and possibly avoid an accident.

Wendy: China

At work we have an open plan office which can get rather noisy and distracting, so I use headphones to hide from the noise and to help myself focus. Ideally, music that isn't too hectic. Something relaxing but that can still drown out everything else around me.

Jahangir: Bangladesh

As with the observations raised concerning background sound, there is a notable degree of diversity in the above responses. Of course, the soundscape of an individual's home is possibly as exceptional and individualised as any single soundscape could be. Consequently, the connections people draw between themselves and the sound objects and events within their soundscape, both by attribution of meaning and extraction of information, will arguably be even more diverse than the soundscapes themselves. Most of us will have comparable appliances and furnishings. The broad architectural structure of our homes will be mostly similar. Despite this, people still utilise the sounds within these environments in dramatically different ways, from deciphering the precise meal their neighbour is preparing, to knowing who is currently taking too long in the bathroom.

Focussing on one particular observation, sound and cooking have seen various pieces of fascinating research in recent years. Arboleda (2019) discovered that increasing exposure to prolonged sizzling sounds at a hamburger restaurant dramatically boosted customers' perceptions of the food and the restaurant as gourmet. Kojima and colleagues (2015) applied principles of sound and cooking to robotics, developing a model of 'auditory scene understanding' based on sounds of the kitchen (the robot could sonically distinguish stirring, slicing, peeling, etc.) to drive procedural behaviours in a cooking support robot. Most closely related, however, is a 2015 review by Harris that observes how deeply underrepresented auditory awareness is in culinary skill and proceeds to share numerous instances of "sonic instruction in recipes". These include 'the knock', a tap of the knuckles to the bottom crust of a bread loaf in search of a sufficiently hollow sound, indicative of the bread being fully baked. A little more visceral is 'the crack', a technique in roasting duck that instructs the willing chef to place the duck breast-side down and push hard on the backbone until a distinct crack sounds. Lastly, both of Laksha's observations on the pop of the mustard seeds and the sizzle of the oil are directly referenced in Harris' review as important culinary sonic skills. Laksha may have missed her calling as an international gourmet chef.

The last two comments raise another curious observation regarding situational awareness, or rather, the absence of situational awareness, through sound. Both point to headphones as a technological factor, highlighting situations in which awareness is highly important and should not be sonically obscured, but also circumstances in which awareness is wholly unwelcome, and instead *should* be obscured.

Challenges with sound

Though relatively rare amongst all the responses, a select few contributors chose to discuss ways in which either themselves or people close to them, managed what could be described as more atypical challenges with sound. I was genuinely considering leaving out this particular set of commentaries. This was not because they were not worthy of including, but precisely because they were such important points to make. My concern was that there was not the scope to explore them properly within the constraints of the book. As such, I make no claim that the challenges of sound as a theme are explored comprehensively here, but I do insist that the comments below make for fascinating reading.

My partner has traits of obsessive-compulsive disorder and is an extremely light sleeper. When they're sleeping, every sound I make just feels so loud. You really do become aware of how much sound there is around the home. Even turning off power sockets will wake him. We actually have a plan for various night-time routine actions because of the sound. For example, I don't turn the light on when I go to the bathroom because it turns the extractor fan on. Also, I'm careful to keep sound separated between the rooms. So, if I go into the kitchen in the evening, I'll shut the door behind me to minimise the sound. In the lounge, I won't use the subwoofer for the television and will keep the volume, especially the bass, very low. Most of these sounds don't even cross my mind during the day because I don't need to worry about controlling them, but everything feels much louder and more prominent when I need to be quiet so my partner can sleep.

Laksha: Sri Lanka, UK

I feel sound very much as a response in my own body. I can get physically annoyed by sounds that I don't expect. It's called hyperacusis and has a common association with tinnitus, especially with loud or strong sound. It feels like someone is hitting you. A slamming door is a good example of a sound that can trigger this feeling. I actually like drums, but I have to be very careful to wear over-ear protection, otherwise it genuinely hurts. The pain is inside my ears but it feels like an impact from the side. I can feel it in my bones as well. I need to avoid certain places that produce certain sounds, industrial areas for instance. I could not live near a factory because of the sound. I also have to avoid street works because of the sounds of the jackhammer and I always cover my ears quickly if an ambulance or emergency vehicle is passing by.

Angélica: Spain

If I hear specific eating sounds, such as an apple or crisps, I find it very irritating. It's not that the sound is too strong or too loud. I actually looked it up and discovered the condition is called misophonia. With misophonia, certain sounds can make you feel very angry, upset, or disgusted. I sometimes think I may have that, though that is a self-diagnosis. I remember travelling on the train one day and the person next to me started eating an apple. In that moment, I literally had to get up and change seats. I just couldn't stand it. All the other sounds and everything else fades away. I cannot think and all I can focus on is that sound of the person eating, which is the one thing I don't want to do. Since that day on the train, I actually make sure to always bring headphones and music with me, so that I can drown out the uncomfortable noises and have control over the sound.

Anonymous: Greece

In research, hyperacusis describes an exaggerated response to otherwise non-threatening sounds, at a loudness that would be comfortable to most people, whilst misophonia refers to a more generalised dislike of sound. Today, hyperacusis is classed as a rare condition (Coey & De Jesus 2021) and, as Angélica stated, has a common association with tinnitus, but also autism (Danesh et al. 2021). In a review of multiple clinical studies by Baguley (2003), a little under half of individuals with tinnitus also experienced hyperacusis, whilst those with hyperacusis were extremely likely (89% of clinical patients) to also be living with tinnitus. Baguley also observes that in most cases, no underlying cause of hyperacusis can be established, the author observing that the patient population is likely to be otherwise heterogenous. In the many years since this work, these observations remain largely unchallenged. For individuals suffering from the condition, Baguley's review closes by arguing against what tends to be patients' first instinct – to invest in earplugs/defenders or find other ways to reduce their exposure to sound. Baguley asserts that doing so has been shown to exacerbate the condition, a result that makes sense if we consider acclimatisation effects more broadly (such as those discussed in Chapter 5). Instead, the opposite approach is advocated, using (for example) binaural recordings as part of exposure therapy with the aim of gradual desensitisation – though evidence of the effectiveness of treatments for hyperacusis still remains preliminary (e.g. Zuschlag & Leventhal 2021). At present, the most recent guidance on the management of hyperacusis comes from the American National Institute of Health and includes several strategies overlapping with managing tinnitus, such as cognitive behavioural therapy, and sound-generator therapies (Coey & De Jesus 2021).

What is your relationship with sound within your home?

In his review of sound in relation to culture and history, Mansell (2018) cites an upsetting piece of research from 1932 that reported on several tenement blocks in Manchester. Whilst little if anything in the descriptions of life in such buildings was positive, Mansell observes that it was matters of sound that made life for these unfortunate tenants so truly unbearable. The non-existent sound insulation between units dictated a persistent sonic onslaught, depriving tenants of all privacy and even creating a sensory blunting effect, as the minds of the tenants seemingly 'muted' all sensation. Their minds defensively shut out the whole world to make the constant sound bearable. In this circumstance, life with sound took away many people's ability to feel anything at all.

Outside sound sources and architectural effects

Based on what could be ascertained from the discussions, I am happy to report that none of our contributors appear to have been suffering with the above challenges, but they most certainly did, on frequent occasions, bring up matters of the architectural features of their homes, and the impact that had on their relationship with sound.

In the UK, you cannot normally hear sound from people living next door. In Japan, the legal rule for making noise is different. It's not as strict, but the biggest contributor to the sound from neighbours is the building's material structure. Traditional Japanese houses are wooden buildings with wattle and daub walls, which are less soundproof than brick houses. This gives a lot less insulation and sound travels between rooms and homes more easily.

Shoko: Japan

There's definitely a difference between Malaysia and the UK in terms of sound insulation. In Malaysia we mostly only have single-glazed windows because the climate doesn't justify the use of double-glazing. This does mean that it is more difficult to shut out any undesirable exterior sounds.

Anonymous: Malaysia

There's almost no carpeting in homes throughout Costa Rica. You might have a few rugs, but a lot of houses have wooden floors, which have really nice creaks. In more modern apartment buildings, you'll be much more likely to have tiles. In these spaces the sound really travels, so you'll hear neighbours' chairs scraping across the floor. The sound of heels on the tile can also be really loud.

Arturo: Costa Rica

It is quite uncommon in Bangladesh for homes to have glass-pane windows. So, in terms of sound, you effectively have a hole in the wall with no barrier. At night in the city, most of the street sounds continue. To get to sleep, you have to get used to the level of sound because you simply can't shut it out. After a while, you do adjust to this, but at first it can be very difficult. It also means that any concerns about internal sounds from neighbours is completely trivial in Dhaka because any of those sounds are completely inaudible over the external sounds coming in the through the open windows. It's not unheard of for some people to have glass windows and insulation in Bangladesh, though that's largely reserved for the very wealthy.

Jahangir: Bangladesh, UK

When I first moved into my home it was very quiet. There was jungle space around, so it was very green. Then they built a highway through that green space, and of course that construction took several years where the sounds outside were very different. Once it was complete, the number of cars increased over the years and the sounds became louder. I would actually call it noise now. From my bedroom, I can always hear this traffic on the highway. Now, I have the option of changing my bedroom to the other side of the house, but five times per day, the Muslim Adhan can be heard clearly from that side. The Muslim places of worship can vary in size, but they all have loudspeaker systems affixed to their towers that announce the Call to Prayer. As the years have passed, the volume from these speakers has clearly gone up.

Anonymous: Malaysia

Jordan has a mild climate and much of the interventions that people make in their homes are related to temperature. During the summer, you would expect most homes to completely remove their carpets, leaving all the flooring as tiles. They would even remove the curtains or replace them with much lighter fabrics. People in Jordan are unlikely to use air conditioning. Instead, they will rely on more passive ventilation. In winter, they will put back the carpets and heavy curtains. This really affects the sound. In the summer, the sound is a lot more echoing, whilst in the winter it is more muted. The homes in Jordan also have much higher ceilings compared to the UK so, particularly in the summer, there is a distinct echo on all the sounds in the home. It's quite an empty feeling in the living spaces.

Dana: Jordan

Differences in material are particularly prominent features here, applicable mainly to walls and floors, with the former affording widely varied levels of sound insulation and the latter, equally varied levels of sound creation. Contributors also draw connections between the build of their homes and other cultural factors, referencing sounds relevant to religion, differences in urban and rural soundscapes, and attitudes towards noise.

As one might expect, climate can influence architectural factors quite significantly, with changes in a persons' home build, in turn, considerably affecting their daily domestic soundscape. The implication is that, as climate can be so very different around the world, so too must be the construction of the home, and therefore the most meaningful soundscape (at least in terms of hours of exposure) in a person's life must be dramatically different around the world. It also raises a largely unreported eventuality of climate change. Just as countries such as Malaysia cannot justify the auditory privacy-affordance of double glazing, and Bangladesh appreciates little value in glass windows altogether, what would be the societal impact for a nation previously used to the sound insulation if it were someday expected to shift to single (or perhaps even zero) glazing as the new climate simply didn't justify those additional panes of glass?

Sources of sound from inside the home

Whilst the previous section dealt with our domestic experience of sound through the boundaries of the home's construction, we now turn our attention to how sound generated by way of objects and events within the home can contribute to this domestic soundscape.

> Our home is generally quiet. One thing that can be notably loud, but is okay in the context of our home, is the Adhan. Unlike other places, where the call is broadcast over loudspeakers outside, here in Portsmouth we have a radio in the home that is directly connected to our mosque. The Call to Prayer is the same but is broadcast live, directly into our home. Most Muslim households in Portsmouth have this type of radio. For us, the radio sits in our dining room. I don't think there are any more loudspeaker broadcasts of the Call to Prayer. Most likely due to historical noise complaints.
>
> Jahangir: Bangladesh, UK

In the summer, across most parts of Iran, you do notice a change in the sound because people start using their air conditioners. You get this routine background sound of the fan everywhere, and the windows will stay closed. In most modern houses, the windows are double-glazed, so this combined with the air conditioners, and you can't hear any noise from outside. Any sound you do hear is from inside the house; from your appliances, so the air conditioning, the fridge turning on and off. These are the normal noises you will hear.

Mehrdad: Iran, UK

There is a lot of music in my house. I have keyboards and a guitar, that sort of thing. There's the fire alarm in my house which really irritates me because it is too loud, and it's too sensitive. I think the water I use must be too hot, because every time I open the bathroom door, the alarm kicks off. Every single time this happens, and it happens at night.

Anonymous: Kenya, UK

One sound that comes to mind is the ticking of clocks. In the lounge of my home for example, even if it is otherwise silent, there is always the sound of the clock ticking away. For me, it's a calming presence, but at the same time it can become irritating when you fixate on it. I've had friends visit who had to ask me to take the clock out, they found it so irritating.

Laksha: Sri Lanka/UK

Across Jordan, the structure of living in the home changes across the year. During the summer, families will be more likely to spread out across all the rooms of the house. During the winter, they will group together and sleep in the same room; typically, the main living room. This is partly because it is more economical in terms of heating costs, but families in Jordan genuinely enjoy this change towards living in a more social space. This of course affects the sound as rather than being more distributed through the home, the activity is concentrated and lively in a smaller space, with unused rooms cold and silent. The flooring of homes in Jordan also present a distinctive sound in terms of cleaning the home. Homes have drainage throughout the building that allows people to clean their tile floors quickly with buckets of water. So, rather than a vacuum cleaner, you'll hear the sound of water splashing across the tiles and draining away at the edges.

Dana: Jordan

Various beeps, clicks, ticks, and pings make frequent appearances throughout the above commentaries, mostly relevant to generic household appliances and furnishings. Cultural factors such as religion again contribute to some of the specificity between these responses, as does geographical features such as climate, albeit indirectly. It is also particularly interesting to observe the reflections many of the contributors make regarding the meaning of these sounds to them, describing the internal sounds of the home more as points of interaction than static objects. The sounds can evoke various emotions, irritation and frustration being two repeat offenders in this context, but there is also much positivity. The last of these comments is a particularly striking example of this. The winter months drive a concentration of human sounds within the home that is experienced as a joyous circumstance, one that is missed when the temperature rises and the use of the space becomes more separate.

Sound, stress, anxiety, and relaxation in the home

My home is calm and quiet, just how I like it. I may need to leave the room when *EastEnders* comes on the television.

Mehrdad: Iran, UK

Where I live, it is a semi-detached house and is mostly very quiet. What I find is that, because it is so quiet, if there are any sudden noises it makes me a bit stressed. If I hear a cat in my garden or a seagull on my roof, these sounds are unusual, and it gives me this sense of "what is happening?". In Shanghai, it's very different. A high amount of sound is quite normal, whether it's my neighbours making noise, or the sounds of all the cars, or the sounds of children playing and laughing. The sound is like flowing water. Hearing slightly unusual sounds there, I don't get that same sense of anxiety.

Wendy: China, UK

In terms of stressful sound, the first thing that comes to mind is the noise from the road. That's a predominant and invasive noise. Then there's sounds like microwave beeping and the notifications on my phone. I have disabled the vast majority of my notifications on my phone because

I hate these sounds. I describe myself as a mono-tasking person and this is why I hate such notification sounds because they distract me from what I am currently doing. The doorbell is similar but not so much of a problem, partly because it is much rarer, but also because it is the expression of a living being. Somehow, having an 'in the flesh' postman ringing the bell is nowhere near as irritating as some bloody notification for an automated message telling me of something I should have done yesterday.

I sometimes use YouTube and play music literally entitled 'relaxing music'. It's interesting, for me it can remove stress and anxiety to some extent, but my partner absolutely hates it.

Anonymous: Italy

Outside my home, I can sometimes hear neighbours talking loudly which can make me feel a little uncomfortable. They are not fighting or shouting but, to me, their voices seem to naturally sound louder and more aggressive. The intensity of the tone feels very negative and I can find it disturbing.

When I first moved to where I am now, I noticed a regular sound at around 5am every morning that sounded like crying and I found it quite disturbing. I knew my neighbours had an older relative living with them, so the sound did make me feel worried. After a few days, I noticed more of a regular melody in the sound and suddenly recognised it as the Call to Prayer. This was a real relief and, once this happened, the sound immediately stopped making me feel uncomfortable.

Smita: India

It is quite apparent from the above, that our respondents largely interpreted this question on sound in the home with regards to negative associations, with only one instance raised of sound being used more positively for relaxation (and even then, the sound in question was wonderfully depicted as completely polarising between two people). In an observation that resonates with several others made previously throughout this book, a number of respondents described processes of acclimatisation to previously stress-inducing and noisy soundscapes. Others, conversely, point to circumstances in which prolonged exposure to a quiet environment leads to acute spikes of anxiety when sudden noises punctuate the otherwise peaceful sound. A repeated association is also drawn between unanticipated or unknown sound and feelings of stress or anxiety. Misattribution of meaning to a sound, or simply not immediately being able to identify a source, is also associated with discomfort. Lastly, respondents also provided examples of when they have actively changed their environment in response to its sound. From disabling their phone notifications to leaving the room to avoid *EastEnders*, they seek to control aspects of their personal soundscape to improve their experience within it. I used to do exactly the same when *Strictly Come Dancing* invasively pirouetted its way into my home on a Saturday evening.

Good and bad sounds in an everyday context

As seen in previous chapters, culture is largely synonymous with 'place' and the unique soundscapes in which we are immersed everyday must surely be influencing how we perceive sound.

Pre-interview observations

In advance of the interviews, and with a view to unpicking initial responses, contributors were asked in pre-interview questionnaires to have a go at R.M. Schafer's *seismographic* listening exercise (1970). Described previously in Chapter 2, this little task required the contributor to take 5 minutes to simply listen to the world around them, then to describe that place in terms of the two fundamental acoustic values of pitch and loudness, but also the decidedly more subjective value of beauty. Some of the results from this section of the questionnaire are presented over the next couple of pages (Tables 6.1 and 6.2).

Table 6.1 Pre-interview questionnaire responses: identify the loudest and quietest sounds

What is the loudest sound you can hear?	*What is the quietest sound you can hear?*
Streamer's voice (listening to YouTube)	Speakers going to standby
Rubbish collection truck	River, flowing water
The quietness	Snow falling
Mosque speakers for the daily prayer	Paper rustling on the table
Music and loud voices in nearby restaurant	Sound from the sea waves
The bang from truck and traffic sound	The tweet of a bird
Ambulance/police car siren	Voices on the pavement beneath my window
Drilling machine, grass cutting machine	Birds
Clattering in the kitchen	TV noise in the adjacent living room
People talking	Keys on the keyboard
The fan from my computer	Fridge
Typing	Humming of computers
Street sellers	Music
Car horn sound	Waves of the sea
Petrol lawnmower	Laptop fan
Plates scratching in kitchen	The buzz inside my head
Building works with glass smashing across the road	Cat meowing
Air conditioning	Electrical hum from lights/leads/wires
My own voice	Voice of a very shy student
Traffic noise, thunder, washing machine	PC sound, fridge
A motorbike zooming past	Clock ticking
TV	Computer fan
The noise of TVs	Family members talking to each other

Table 6.2 Pre-interview questionnaire responses: identify the most and least
beautiful sounds

What is the most beautiful sound you can hear?	*What is the least beautiful sound you can hear?*
The cat meowing	The grill
Music	Eating sounds
When my boy finishes nursery call me	Phone vibration
Music of the game	PC fan
Birds chirping	A car horn
Someone cooking	The mosque call to prayer
Aeolian bells	Construction noise
The song of a goldfinch	Nails scratching a blackboard
soft, low-pitch white noise from trucks	Ambulance/police sirens passing
The Tanpura	When people shout
A windchime	Creaking chair
Birds singing	Sound of crows
Rhythmic typing	Coughing
Family laughing	Passing street sellers
Cello	Alarm
A woman's voice	Clock
I would describe none of the sounds as 'beautiful'	Electric saw
My baby daughter breathing	My brother's last breath
Two people coincidentally clicking their PC mice simultaneously	Chair squeaking
A friendly voice	A cold voice
Instruments (piano, cello, violin)	Traffic in the crossing
Wind	Drilling
Birds	Passing truck
Children making noise (laughing, talking) while they play	A door closing

The more dispassionate responses above broadly identify various sound objects and events, presenting relatively few surprises. The noises of construction and maintenance works, domestic appliances, traffic, and also alarms
and sirens are prominent features of loud sounds. By contrast, the list of quiet
sounds more commonly features animals, natural environment events, and
occluded sound objects that are in some way obscured from the listener by
a physical barrier such as a wall, floor, or door. It's also notable that many
of the sounds identified as loud are also sudden, or at least non-continuous
(thunder, glass smashing, passing motorbike, etc.). Many 'quieter' sounds we
would expect to be persistent within the environment (flowing water, waves,
clock ticking, fridge hum, etc.). One curious feature of these responses is
the appearance of certain items within both lists. The whirring of computer
fans, human voices, music, and typing on a keyboard all feature at least once
on both lists. Next, in what I would have anticipated to be the more varied,
personal, and even disagreeable list, the most and least beautiful sounds identified below did raise a couple of surprises.

The first surprise was more of a methodological issue, possibly due to the room for misinterpreting my wording of the question; particularly considering respondents for whom English was a second, less fluent language. Looking over several of the responses above, I feel at least a couple of contributors had misinterpreted the question to mean "*what is the most beautiful sound?*" as opposed to the intended "*what is the most beautiful sound that you can hear, at this particular moment in time?*". Of course, I don't actually have the evidence to know for certain. Maybe one of the respondents really was in a place within which they could in the very same moment hear nails scratching across a blackboard as a Goldfinch sings merrily. I'm inclined to be sceptical. The second surprise, not to be too facetious, was that there was no further surprise. Various objects and events identified within the most and least beautiful lists reflected my expectations, as I predict they did yours. Particular musical instruments or non-musical objects create sound that has musical qualities, birdsong, vocalisations of people to which the listener has a significant bond, and auditory expressions of positive emotion capture almost every item on the list of beautiful sounds. Conversely, the invasive, sudden, and almost certainly distracting sounds of traffic, construction, and sirens dominate the least beautiful list; also demonstrating a significant overlap between this and the perceived loudest list.

Considering research that explores beautiful sound, several instances focus upon a particular sound object or event that polarises opinion. In sport, for example, Surujlal and Mafini (2011) observe the deep opinion split regarding the Vuvuzela, between beautiful sound (evocative of cultural identity and camaraderie) and nagging noise (distracting and invasive). Experimental data acquired by Delplanque and colleagues (2019) evidenced an 'inverted-U' relationship between complexity and preference towards a sound, essentially suggesting that there is a broad human-level consensus on what constitutes a moderately complex sound and that such a sound is more likely to be perceived as beautiful. A 2013 competition run by *BeautifulNow* judged the 'most beautiful sound in the world' to be the soundscape capture of a Malaysian swamp at dusk[1]. Further popular-facing sources are rather fond of putting together compilations of sounds deemed to be universally beautiful or satisfying. Repeat offenders on such lists include the bite of an apple, planes flying overhead, the jangling of coins in a pocket, the rush of air when lifting the cap from a beer bottle, and of course, birdsong.

As Hartshorne (1968) posits: "Birdsong is a fine illustration of beauty of an un-profound sort (by human standards). For human ears it is pretty rather than sublime. Sometimes it seems a bit chaotic, sometimes rather monotonous, but seldom hopelessly either" (p.311). On the behavioural basis that more recordings of particular birdsongs imply a greater perception of beauty in that sound, Blackburn and colleagues (2014) observed that the complexity of a bird's repertoire and the extent to which the species was distributed across the globe were both strongly related to perceived beauty. So, if a bird is well travelled and has a solid compilation of hits, we're apparently more likely to buy their album.

Returning very briefly to the questionnaire responses, and on a personal note, I must protest in defence of the humble crow and the glorious grill; their respective cawing and hiss being two particular sounds that I enjoy very much.

The good, the bad, and the "well, it depends..."

Expanding upon their answers in the pre-interview questionnaire, the following is a selection of interview responses to the direct question: "*what comes to mind when you think of 'good sound'?*"

> One of my favourite sounds is my son calling out to me at nursery pickup. It's a really familiar sound. Where there's been a gap, a distance of time where you've been away from the voice and have had time to miss it, then the voice returns, it brings up a lot of emotion. If it were a five-minute gap, the voice wouldn't have the same effect. But after a few hours it evokes that sense of missing and the positive emotion when you hear the call again.
>
> Si: China, UK

> I immediately think of playing and the sounds of children. The word 'noisy' certainly comes to mind when I think of children, though, at the same time, I do like being outdoors and hearing the sound of children playing. It feels almost torturous to put a child in a confined space, and that's when the sounds of children are noisy or indicate things that the adults need to look out for. But the sound of them playing outside, their laughter, has a sense of excitement that almost makes me feel like I should be playing more. More jumping up and down. It's quite nostalgic. These sounds make me think of my own childhood. It's a cacophony of sound, but one that sounds absolutely joyful.
>
> Anonymous: Kenya

> I have a memory of always liking the sound of starlings. Through my teenage years, I remember that, watching films and television, I didn't like it when certain sounds were changed. I enjoyed a lot of science fiction sound in which they would use very particular sounds for certain far out fictional things. A good example of this is the sound of the Photon Torpedoes in the early *Star Trek* motion pictures, which was changed in the later movies. They really shortened the sound.
>
> Stafford, UK

> One of my favourite sounds is definitely when it rains, and you have that lovely sound where you cannot hear anything else other than the rain just pouring down. I absolutely love that sound. I actually downloaded an app that had various rainforest sounds. It was okay, but it just wasn't the same. It bothered me that it wasn't real, that I wasn't hearing it naturally.
>
> Anonymous: Norway

Music is always pleasant, whatever style is playing, as long as it's at the right volume. When I'm in the office, I will always prefer to have some kind of background music playing. When I'm working alone, I like to play the radio from my computer. I also enjoy the sounds of nature. The other day I stopped because I could hear birds and wanted to spot where they were. Natural sounds really give me a sense of 'wow'. The voices of my friends and my family, my son, are very positive – unless I feel they're too loud, then they're negative!

Anonymous: Greece

Resonating with a mirror-reflection of unknown sound evoking anxiety, several respondents characterise that which is familiar as consistently good sound. Children also manage to feature in our nice list, and this is strongly connected to the affective state they are expressing through sound. A happy child appears very much to be a good-sounding child. Several sound events identified as beautiful in the questionnaire responses also appear prominently in the discussions. These include birdsong, rainfall, and music. It is worth noting here that the particular sound of the *Star Trek* photon torpedo was an observation that emerged from a discussion but would have been highly unlikely to appear in a questionnaire, as explaining the much-needed context would have been difficult. This is a wonderful sound to include, however, both in itself (take a listen[2]) and as an example of a single sound becoming embedded in memory to the extent that it becomes a permanent feature of a person's cultural history and, consequently, a point of ongoing potential influence upon their perspectives, interpretations, and interactions with sound. That represents the good, but what about the bad?

When I think about noise, I actually think about family. I appreciate that's not what you might expect me to identify as a negative sound. I come from a larger family. I have a lot of extended family and they collectively generate a lot of noise. When I was younger, I would always have something to block my ears when around family because of the noise, especially when I was reading.

Dana: Jordan, United Arab Emirates

I dislike the sound of my phone vibrating. It's not actually the sound itself that annoys me, but the fact that I put the phone on vibrate specifically because I didn't want to be disturbed. I can't switch the phone off in case a call is an emergency, but the vibration grabs my attention and so it is still a disturbance. Often, it's someone who rings through many times, so presumably they know that you're busy but keep ringing anyway. I've also found that the meaning of a ringing phone has changed with my age. When I was younger, a ringing phone might mean a chat with friends, an invitation to something, or some good news. Now it's just people trying to sell me broadband.

Si: China, UK

I hate the car horn... A lot of my colleagues regularly listen to music through headphones, on the way into work or in the office. I actually hate listening to music when I'm trying to walk anywhere. I like to be aware of my surroundings. Even if I'm out running, I can't do that to music. It's not out of fear or anything like that, it's more that it feels unnatural hearing music when I'm outside. I like to take in the sound of the environment as it is. In the office, I always make use of the silent or vibrate options for my phone as it can grate on me when it rings. I don't feel any kind of aversion to the 'call to action' of a ringing phone, it's the sound itself that I don't like. Sound is fine in the background, but intrusive, particularly artificial sounds of electrical devices such as the television and radio, I definitely dislike.

Anonymous: Norway

There are certain noises that will certainly put my teeth on edge. Whether it's the classic chalk on a chalk board, or a pneumatic drill out on the street, those sounds just go right through me. For me though, truly bad sound is silence. I'm much more comfortable using radio or television sound to recreate the sense of another person in the room, rather than being alone in silence.

Alyson: UK

I'm uncomfortable with complete silence, but I'm equally uncomfortable with intrusive road noise. Silence gives me an impression of total loneliness and I find that extremely unpleasant.

Anonymous: Italy

Our 'bad list' of sounds also appears to reliably corroborate our questionnaire responses, at least to some extent. Car horns, vibrating phones, roadworks/construction noise, and general traffic sound are all repeat offenders. Where the discussions add notable richness however is arguably in their capacity to identify *situations* of bad sound rather than simply objects of bad sound. In this context, bad sound is such: (1) when it is distracting from a task; (2) when it demands we attend to it at a time that we are otherwise engaged, or if the required response is simply something that we have no desire to do; or (3) when it is obscuring our sense of presence, potentially limiting our situational awareness or simply taking us out of a moment. Also of note was the difference between respondents on whether it was the acoustic properties of the sound or its perceived meaning that triggered the negative response. In the case of the ringing phone, for example, two people may clearly despise the sound they are receiving, but for two completely different reasons. It is also important to acknowledge the final two comments made above, that describe aversion to silence. The revelation of this in some of the interviews was unsurprising, particularly considering some of the comments made earlier on the positive usage of sound for background ambience, suggesting that whilst only a few contributors directly expressed their discomfort with silence, many others would do the same if pressed, based on their acknowledgement of actively and persistently filling their world with sound every day.

> I think of my positive feelings towards the loudness and chaos of the soundscape in Alexandria, that I appreciate many people may experience as a negative. It highlights how much difference context can have on our perception of sound. Whilst living there, I was, in many ways, in Egypt on an extended vacation. I think that put me in a mindset of being more curious and exploratory. If I had to work 9-5 in that environment, I appreciate that I might have struggled with the sound, but I had the time and the freedom to explore.
>
> Ludo: Italy, Egypt, UK

> Where I live now in London, I feel as though sound is mainly negative. Outside my home there is the sound of sirens and traffic which makes you try to shut the sounds out. I think it's the city life where sound tends to be more intrusive. When I first moved to London, I actually didn't think that I would be able to cope. It's strange, but I've accommodated a lot of the sound over time. There's a highway close by that produces a lot of noise. At first this felt so overpowering, but I don't feel as though I hear it as much these days.
>
> Anonymous: Kenya

> Particularly now I'm working from home, sound is a real distraction. I find it interesting that, to me, the sound is negative because it is loud, yet to my son it is positive because it is loud. He loves loud sounds, especially if he's the one making them. I love rock music but, in that instance, loud sound may be good for me, but not so much for my neighbours. It's all personal and relative. I also found that soft sound, or even no sound at all can feel just as disturbing as a loud sound. We had a new washing machine installed recently which operated extremely quietly, to the point I didn't believe that it was actually working at all.
>
> Si: China, UK

Acclimatisation returns in the final comments of this section. The notion of extraneous sound in relation to task also resurfaces, but here it is observed for the scale of time can extend to periods of weeks, months, or even years. A chaotic and arbitrarily attention-grabbing soundscape may be highly suitable if a person's reasoning for being in that environment is more exploratory and dispersed. In such an instance there effectively is no extraneous sound, with everything in the soundscape being potentially fit for purpose. However, if the priority is to attend to something more focussed, any sounds to which we attribute meaning that falls outside of this focus, become extraneous distractions we will likely seek to shut out. The last of these comments also reminds us of interpersonal perspective. Here, in the same moment, with two people exposed to the exact same soundwave, object, or event, for one it is unquestionably a good sound, whilst for the other, it is unmistakeably bad.

The etiquette of sound

Closing this chapter, we now round off our consideration of everyday sound with a brief look at how sound features in social interaction, specifically how we are (or are not as the case may be) considerate of others in our production of sound. What is the etiquette of sound, and does its meaning differ significantly between people and place?

Sound etiquette in domestic life

We begin with a little matter of some personal significance. In 2019, I lived in an apartment building on the Southsea seafront. When I was shown around the flat, I distinctly remember the letting agent overtly bragging that the soundproofing was outstanding, because the building had previously been a hotel and much had been invested in the insulation. A few weeks in and I discovered that whatever 'hotel-grade' insulation they may have used could not withstand the herculean might of Drum and Bass. Every Friday and Saturday night, the neighbours below me would *begin* their festivities at around 11 pm. They didn't finish up until the following day, around a time at which you might wonder if it is too late for breakfast or too early for lunch. Then began that wonderful loop of British passive-aggressiveness in which I asked them to turn the music down, to which they would immediately oblige, but then resume their normal volume and routine the next night, each time I asked them they responded as if it was the first request I had given. Pretty quickly it became too socially awkward to say anything. So, whilst I try to put my bias to one side, we return to our contributors and their experiences.

> Across New Zealand you have a lot of different communities and cultures, and they all enjoy music and dancing. In terms of social behaviour and sound, say with neighbours, I think there's just a general rule of common sense. You wouldn't expect to hear loud drum and bass at midnight for example, but generally there isn't a social restriction on sound.
>
> Anonymous: New Zealand

> For me, when someone produces any kind of sound that is loud enough for other people to hear it when they don't need to, that's being rude. Someone having a phone conversation right outside my bedroom door at 8am; that's a good example of me hearing a sound that I shouldn't need to hear. I think the time of day that making sound is acceptable is very personal. I guess it mainly depends on when you want to go to sleep. I go to sleep around midnight, so neighbours making noise until then is

personally okay, but not after that time. I think the etiquette of sound is dependent on an individual's routine and lifestyle.

Jahangir: Bangladesh, UK

When people get really involved in what they're doing, they don't notice things in their periphery. We shut out sounds. If my partner is engaged on the computer for example and I appear without making any really noticeable sound, his reaction is often shock. I've found myself intentionally making loud sounds before I enter the room to announce myself as I'm coming in. I empathise with my partner because I have had moments where housemates have silently appeared, and I've jumped out of my skin. I think I treat people in the way I want to be treated. I always try to be respectful with the sound that I make as I know I'd get really irritated by people being thoughtless with their sound. My partner often has the television on whilst he's on his phone. I find that quite irritating but, actually, I do understand why he does it, because otherwise you're on your phone in complete silence, which is uncomfortable. It can be awkward, because I'm not sure if he's interested in the programme or if I can change the channel.

Laksha: Sri Lanka, UK

Living in the city and in large apartment complexes, most people will be part of an online social group using a platform like WhatsApp to communicate and keep up to date with issues around the building. So, if someone was making a lot of noise in one apartment, it's not unusual for that to be a topic of discussion online. Generally, people don't want to be the subject of that kind of discussion so they will usually try to keep it quiet. I think that there's generally a recognition of the importance in being respectful about the sound you make around the people who can hear you.

Andreina: Venezuela

I think the etiquette towards sound is very relaxed in Spain. If you wanted to listen to loud music at home, you wouldn't really ask permission or worry about it disturbing anyone else, but at the same time, if you were disturbing someone, they be comfortable telling you right away. There's a general agreement over what time of day it is appropriate for making loud sound. If people were playing loud music at midnight, you'd absolutely have the right to tell them to lower the volume immediately. I wouldn't feel uncomfortable telling someone in Spain to be quiet if I felt they were being inappropriately loud. It wouldn't matter if they were friends or strangers.

Lucia: Spain

The seasonal changes in Jordan definitely affect the social aspect of sound, as the summer means people will have their insulation taken down and their windows and doors open. The sounds all pass between homes. People in Jordan actually prefer this. As winter approaches and people begin shutting windows and doors, they can start dreading winter because of this increased isolation between home and community. In Jordan there is a sense that if you're setting personal boundaries, then you're being rude in some way. In the more mixed, multi-ethnic communities, especially in the refugee camps, you have almost no personal boundaries of sound, but the people of these communities will often talk about how much they value this. It gives them a sense of community and makes them feel as though they are almost family with their neighbours.

There is still a sense of social responsibility in terms of sound in Jordan. One particular example would be if a neighbour had lost someone and was grieving, then you would not play loud music during that time out of respect. If you were celebrating a separate event during that same time, it would be normal for you to ask permission of your neighbour before you made a lot of noise as part of your celebration. There's also a general sense of social responsibility in responding to certain sounds, that indicate violence for example.

Dana: Jordan

I think there's a pretty broad range of attitudes to being polite with sound. It's unlikely people will complain about sound unless it's very loud or at a really unsociable hour. Even then, they wouldn't normally be confrontational. They'd be more likely to log a complaint to, say security, who would contact the person making the noise. We have certain rules in our apartment building and in general for labour work. So, unless its important government work or emergency services, there are rules for when the labour can be carried out. At the same time, Mexican people are fond of parties so you should expect to hear music and the other sounds of a party pretty frequently. These can go on late into the night. You could make a complaint about this but it's not likely anyone would respond. The police won't come out for a noise complaint from a party.

Meni: Mexico

In our family, there was a distinct etiquette around dinner. We would always eat at the table and there was a requirement for having the table first properly laid, having bread buttered, then towards the end, not being able to leave the table until everyone else had finished eating. There was no television or radio allowed whilst at the table. There were no computers, phones, or other devices. There is of course a sound aspect to this. I remember one of my childhood friends used to have regular discussions with her father over dinner. There would be no distractions or background noises. I drew a connection between this and why my friend did so well at school. I actually did better at school myself by listening to the way that they talked together and enjoying these conversations.

Alyson: UK

Whilst in many countries, there are formalised noise-abatement structures in place that aim to minimise noise disturbance, this did not feature in the discussions and contributors predominantly pointed to less formal, common-sense rules. There is some distinction between several respondents with regards to raising complaints, with some insisting no discomfort whatsoever in marching straight to the source of the problem and making their feelings abundantly clear, whilst others describe a more indirect route to resolution. The usage of social media within this context was certainly something I found personally surprising, but it does appear in many ways an elegant and informal (if maybe a little passive-aggressive) solution to managing noise, especially in a multi-unit apartment complex. As with some of the other topics discussed throughout this chapter, perspectives on what constitutes antisocial sound can vary depending on the receiver, with once again task being identified as a significant variable. If the task requires a disruption-free environment, then a persistent, aggressive sound becomes far more invasive and feels more disrespectful. If the task itself does not require a quiet space, the interpretation of the extraneous sound will be different – if it is attended to at all. It bears emphasis that not all contributors even perceived loud music from neighbours in a remotely negative context, instead of describing a sonically porous environment that was embraced by the community and represented a greater closeness between large groups of people. In such a culture, the description given presents an impression of implicit trust and respectful responsiveness between people within the community. Rather than building walls both higher and thicker to protect from antisocial behaviour, the priority is that the community inherently discourages such behaviour, thereby reducing the need for such barriers.

In addition to talking about the topic of invasive music, contributors also raised several observations that I was genuinely surprised by and hadn't initially thought to consider. Two wonderful examples of this include the practice of intentionally generating sound as a way of subtly announcing yourself before you enter a room, lest you make someone "jump out of [their] skin", and the absolutely fascinating observation of how sound can be a central aspect in the etiquette of mealtimes.

In developing designs for a remote-activated dining room table lamp, Djajadiningrat and colleagues (2012) were surprised to discover that systems using either clapping, finger snapping, or voice activation were all rejected by participants, specifically because their sound was disruptive, invasive, and deemed highly inappropriate in the sociocultural context of dining. In a comparative review of Eastern and Western table manners, Cooper (1986) identified numerous differences in etiquette that have significant effects on the sound of dining. One may not expect to hear the taps and clicks of a diner dislodging a hunk of gristle from their teeth at an American table but should be surprised if such sounds were not present at a Chinese table. Eastern tradition is also described as favouring the 'young deferring to the old' in order of eating. The effect of this is that sonic differences in how adults and children eat will typically be clearly separated by space (the children's table) or time (the children eat once the adults have finished) whilst, in many Western cultures, all these sounds are thrown in together. In *The Rituals of Dinner* (2015), Visser makes a multitude of references to how sound features in the etiquette of dining. Visser's work illustrates a dramatic range of attitudes towards dining sound, drawing in matters of status, tradition, expectation, and perceived standards to name a few, all differing across both place and time. One example describes minimising drinking during a meal as much as possible, restraining such activity until at least the second course, and even then, being careful only to sip, never to gulp, all for the 'unseemly' sound it would produce. A dramatic opposition to this attitude, one further example emphatically directs diners *never* to stifle an 'involuntary auditory misdemeanour of the body' in the name of courtesy.

Respectful sound in public

Alongside matters of etiquette in sound within the home, I was also curious to ask our contributors their thoughts on broadly the same social sound issues, but now outside of the home. The specifics were left open, and the contributors were free to report on anywhere that came to mind. The main distinction was related to differences in behaviours with sound when people are that bit more exposed, out in the world.

> If you walked into a restaurant in Malaysia you'd find a strong difference of opinion over the acceptable loudness of children, with some feeling that children should be free to make as much noise as they wish, whilst others feel that children, if they are going to be in the restaurant, should effectively be silent. As with many places though, the etiquette around sound will change a great deal depending on the perceived social class of the restaurant. A higher-class restaurant will demand a much lower level of noise for example.
>
> Anonymous: Malaysia

> A few months ago, on the underground metro, there was a new policy that you were now not allowed to play sounds in public whilst taking the metro because it would disrupt other passengers. Many people liked to watch videos on their phones, without headphones, just letting the sound come from the speakers. This policy came in to stop that.
>
> Anonymous: China, Hong Kong

> In Japan there are rules for using not using phones on public transport. Out in the countryside there aren't specific rules for being respectful

with sound, but most people will consider their sound in terms of how it might disturb the wildlife. Though it does depend on the wildlife. Sometimes we will carry a bell as a way of gently alerting and keeping bears away.

Shoko: Japan

I have noticed that some people in Caracas will listen to music on public transport on their phone speakers rather than headphones. That is quite frustrating as they're imposing their music on you. It's not as though it's gentle, classical music they are playing, but more likely Reggaeton which, to me, is a lot more invasive.

Andreina: Venezuela

In many public buildings or public transport, you'll normally see signs around Shanghai discouraging people from being loud and disturbing other people. I think being socially aware with sound has increased over time in China. People have become more educated and socially responsible.

If you were to talk loudly on your phone whilst on a bus, you should expect people to be staring at you, encouraging you to feel embarrassed. If you don't realise that people are staring and continue to talk loudly, it's then likely that someone will actually ask you to lower your voice. Public transport is typically quiet. The trains in China are also expected to be quiet and most people will just read or use headphones. Also, the access to books and mobile phones means that people in public have more things to do that are solitary or personal. Years ago, without these, people just sitting on public transport and staring at each other would be embarrassing so they were compelled to talk. Nowadays there's access to devices, to read or listen, for studying or for work. Talking is just not as necessary.

Wendy: China

On public transport or outside in general in Spain, people expect other people to be loud, to make a lot of noise. If it were something very extreme, you might see someone intervene whilst outside, but it would be more about the other person being rude or offensive in their language rather than the loudness of their voice. Playing loud music, through phone speakers for example, would be a bit more questionable. You still might not see someone intervene in that case, but you would probably see a few people muttering and complaining to themselves.

Lucia: Spain

Norwegians have a reputation for not being overtly and publicly confrontational, but I think that being considerate about sound is normal. You wouldn't start your lawn mower at 8am or be doing your vacuuming after 7pm. There's a general understanding that you don't disturb people as you're going about your business. In Norway we have something called Janteloven that does create an engrained respectful attitude. It encourages the idea that you as an individual are not inherently special. So, in terms of sound, one person wouldn't feel that they could make more noise, or be more disruptive, because they'd appreciate that they were not inherently special or above everyone else.

Anonymous: Norway, UK

Going back to using headphones in the office, these function as a way of making it clear to the people around me that I need to focus and cannot be disturbed. Sometimes my colleagues are having meetings with students and it's essentially none of my business so I feel as though I should not be hearing that conversation. So, the headphones and music function as stopping distraction but also as me trying to give colleagues their privacy.

Jahangir: Bangladesh, UK

In India, I would say that shouting across other people, say on public transport, is socially acceptable. Conversely, I think the sounds of people eating isn't exactly frowned upon, but people generally don't appreciate it. I know that burping in public is perfectly acceptable in many parts of China, but this certainly wouldn't be accepted in India. I personally dislike all the interface sounds of mobile phones in public, though that seems to be something you will hear everywhere and my disliking of it is probably personal.

One other thing that stands out is the sound of religious processions in India. They can occur at times of the day that other cultures would argue to be antisocial. The sound is very loud indeed. There would be lots of people shouting or singing, many amplified with microphones and loudspeakers. There's also huge amounts of percussion. This doesn't mean people aren't disrupted by the sound, but they know that it is a part of the culture and that there would be no point in complaining.

Anonymous: India

In California, most people are fairly considerate with playing sound out loud in public places. However, we aren't as considerate as other countries such as France or Singapore, where I've seen people take considerable measures to try to be silent in public spaces such as metros or busses. Working in mobile games, we have actually gathered data from our games and discovered that certain countries listen to the audio much more than others. For example, Southeast Asia and South American countries listen to the game audio much more than Nordic countries. The majority of players in Nordic countries turn their sound off when playing.

Michael: USA

Here, respondents explore various different environments, from public streets to restaurants, to the workplace, and of course, public transport. Certain cultural features such as class, religion, and tradition are all identified as potential modifiers in the matter of what is or is not acceptable sonic behaviour. Changes over time are also observed, with respondents explaining how various technological changes, such as the proliferation of mobile technology, have encouraged both individualisation and introversion in certain places – the antisocial sound levels diminishing whilst we all retreat to our personal, digital micro-universes. Unlike the discussion on sound etiquette in the home, here more references are made to formal rules and guidelines, particularly on matters such as loudspeaker usage on public transport. Once again, there does appear to be notable variation between parts of the world in how such sound and antisocial behaviour is perceived and managed, with the final comment above revealing how there are now some fascinating technological means of obtaining quantifiable data that could further inform us on different attitudes towards sound in public, all across the world.

Chapter 6 summary: who doesn't like the sound of a crow cawing?

All things considered; this chapter proved itself to be the one that I found most personally intriguing. That statement may be a wee bit voyeuristic, bearing in mind its content is arguably the most intimate as it explores the human relationship with sound in deeply individual and sometimes even vulnerable terms. Taking a few steps back from the substantial volume of content discussed across this chapter, there is clearly a great deal in how we experience and interact with sound in our everyday lives that connects us; that makes us similar, both in the characteristics of our daily environments but also in our attitudes, preferences, and reactions. However, we are most certainly not identical listeners, and there is a richness in the variations between us – some subtle, some substantial. These differences can give us a wider awareness of the world, but they can also offer us potential practical applications if we take the time to consider how the techniques and strategies people are using to positively shape the sonic aspects of their everyday lives could perhaps be applied to our own. If I ever hear Drum and Bass in the middle of the night again, I am absolutely starting a *Don't be an arse with your music* WhatsApp group.

Notes

1 The 2013 "most beautiful sound in the world": https://soundcloud.com/wildambience/empress-cicadas-frogs-kubah.
2 Photon torpedo sound from *Star Trek: The Original Series* https://www.youtube.com/watch?v=ggcOjx8i5ao (accessed 19.02.2022).

References

Arboleda, A. M. (2019). The Tempo and Cooking Sound of a Gourmet Hamburger. *Journal of Food Products Marketing,* 25(5), 566–580.

Baguley, D. M. (2003). Hyperacusis. *Journal of the Royal Society of Medicine,* 96(12), 582–585.

Blackburn, T. M., Su, S., & Cassey, P. (2014). A Potential Metric of the Attractiveness of Bird Song to Humans. *Ethology,* 120(4), 305–312.

Coey, J. G., & De Jesus, O. (2021). Hyperacusis. StatPearls. Online article: https://www.ncbi.nlm. nih.gov/books/NBK557713/ (accessed 01.03.2022).

Cooper, E. (1986). Chinese Table Manners: You Are How You Eat. *Human Organization,* 45(2), 179–184.

Danesh, A. A., Howery, S., Aazh, H., Kaf, W., & Eshraghi, A. A. (2021). Hyperacusis in Autism Spectrum Disorders. *Audiology Research,* 11(4), 547–556.

Delplanque, J., De Loof, E., Janssens, C., & Verguts, T. (2019). The Sound of Beauty: How Complexity Determines Aesthetic Preference. *Acta Psychologica,* 192, 146–152.

Djajadiningrat, T., Geurts, L., & de Bont, J. (2012). Table Manners: The Influence of Context on Gestural Meaning. In: Chen, L. & Feijis, L. (Eds.) *Design and Semantics of Form and Movement* (p. 27). Amsterdam: Philips.

Harris, A. (2015). The Hollow Knock and Other Sounds in Recipes. *Gastronomica,* 15(4), 14–17.

Hartshorne, C. (1968). The Aesthetics of Birdsong. *The Journal of Aesthetics and Art Criticism,* 26(3), 311–315.

Kojima, R., Sugiyama, O., & Nakadai, K. (2015). Scene Understanding Based on Sound and Text Information for a Cooking Support Robot. In: Ali, M., Kwon Y. & Lee, C. (Eds.) *International Conference on Industrial, Engineering and Other Applications of Applied Intelligent Systems* (pp. 665–674). Cham: Springer.

Manchester University Settlement. (1932). *Some Social Aspects of Pre-War Tenements and of Post-War Flats.* Manchester: Manchester University Settlement.

Mansell, J. G. (2018). Ways of Hearing: Sound, Culture and History. In: Bull, M. (Ed.) *The Routledge Companion to Sound Studies* (pp. 343–352). New York: Routledge.

Surujlal, J., & Mafini, C. (2011). Beautiful Sound or Nagging Noise: Public Perception of the Use of the Vuvuzela in Sports. *African Journal for Physical, Health Education, Recreation & Dance,* 17(2), 25–37.

Visser, M. (2015). *The Rituals of Dinner: The Origins, Evolution, Eccentricities, and Meaning of Table Manners.* New York: Open Road Media.

Zuschlag, Z. D., & Leventhal, K. C. (2021). Rapid and Sustained Resolution of Misophonia-Type Hyperacusis with the Selective Serotonin Reuptake Inhibitor Sertraline. *The Primary Care Companion for CNS Disorders,* 23(3), 32719.

7 Sound and professional practice

My home, Southsea. 9th of December 2021. Morning

I'm returning home after exercising to get to my desk at 9 am. As I enter the atrium of my building and shut the door behind me, there is a satisfying clunk as the door shuts. This is followed by a profound silence punctuated only by my footsteps, which I'm glad to be reminded still exist. What is noticeable here is how warming and comforting these sounds are. The rummaging of my coat, the quick zipping of my pocket, and the jingling of my keys as I position them in front of the front door lock. The decisive crunch as the key penetrates the lock before the rhythmic turn and opening of the door that has almost a tearing quality, akin to pulling the packaging tape from a parcel that you're excited to be unboxing. The solid and heavy hit of the door shutting behind me. All these sounds are extremely familiar and represent a return to a known, safe, and comfortable place. The last thing I hear is myself. A contented sigh.

We now reach the last stop of our expedition. Of all our contributors, a sizeable proportion were active professionals who worked with sound across a range of roles. Musicians, sound designers, audio engineers, foley artists, game sound developers, mix engineers, voice-over artists, and directors – the list continues. Many of our contributors identified their profession as a collection of more than one of these roles. To identify them as a collective I use the term 'sound professional' throughout, though will name specific roles where appropriate. Reflecting on the discussions with our sound professionals, two things immediately come to mind. The first is something that they shared with every individual that I was fortunate enough to interview, they were really nice. The enthusiasm they expressed when discussing their work, their humble and unassuming nature, and their leanings towards openness and self-reflection, all meant that I left every interview with a genuine smile on

DOI: 10.4324/9781003178705-8

my face. The second is arguably more unique feature that I observed was that sound professionals *love* to talk about their work.

A small number of contributors whose professional background was not in sound did discuss their relationship with sound in a workplace context, but this was quite a rare occurrence and is a clear opportunity for future investigation. Consequently, we begin this chapter with those few key remarks before progressing to various matters concerning the distinctive relationships between sound and sound professionals. Our discussion asks how working with sound can affect a person's everyday experience of it. We explore how cultural factors feature in professional practice, and we also question the extent to which the sound artefacts being produced are expected to reflect any culturally related requirements. Are there different expectations and standards for sound across the world? Are new technological affordances and connectivity moving us towards a truly 'global sound'?

How does sound feature in your profession?

Research on workplace sound is largely concerned with matters of noise exposure and its association with various negative health effects. Conducted in a typical office environment, a study by Leather and colleagues (2003) revealed that whilst higher ambient noise levels in themselves were not enough to directly raise employees' self-rated job satisfaction or company loyalty, lower ambient noise levels were shown to attenuate the impact of negative 'psychosocial job stress' factors, effectively making a calm sound environment a buffer to workplace stresses. A little more recently, Ma and colleagues (2020) examined the effects of noise within dental surgeries. Their findings evidenced significant links between the loudness and sharpness of sound within the workspace, raises in noise sensitivity, and decreases in job performance. Over in Brazil, a 2019 investigation by Portela and colleagues looked into sound pressure levels experienced by physical education teachers. They found that although the acoustic levels were not objectively high, PE teachers reliably reported high levels of listening discomfort. On closer examination, what appeared to be the root of the problem was hypothesised to be what the researchers called 'noise interference', or what I would call irritating children constantly disrupting my lesson. Outside of sound, research has explored the workplace environment from a more phenomenological (or, experiential) perspective, such as Loder's (2014) review of 'green roof' aesthetics and their impact on daily working life, but it has not greatly explored the role of sound in this context. Indeed, based on the frequency with which contributors *not* identifying as working with sound chose to discuss how sound featured in their daily experience of the workplace heavily implies that, unless it is disruptive noise, most people don't really consider sound at work. Of course, this does not in any way mean that sound is not an important feature at work, more that its effects are largely happening under the surface.

Communication and feedback

As we move into what we now, with cautious optimism, call the post-Covid era of 2022, the role of digital and network technology in workplace communications has found itself expanding quite significantly. This means, amongst many other things, that much in the way of communication is now conducted by way of a digital mediator, namely videoconferencing.

> I realised that when teaching remotely during the pandemic, I didn't really need students to use a camera, but I did need them to use their microphones. Even if they weren't speaking, the buzz of open microphones in the classroom really sparked a sense of interaction. Sound is part of my interface when I teach.
>
> Ludo: Italy, Egypt, UK

Recent studies reviewing videoconferencing technology generally point to consistently high ratings in terms of both video and sound quality, though several observe that ratings for the latter are lower (Archibald et al. 2019), suggesting that further technological improvements in videoconferencing should be sound focussed. For Oeppen and colleagues (2020), consideration of human factors in videoconferencing dictates that any bandwidth limitations in connectivity should always first lead to sacrificing video so that the quality of the sound, both in terms of latency and fidelity, can be maximised. In a similar vein, a 2021 report by Higashiguchi and Shibuya observed a link between silence in a videoconference call and significant increases in participant stress, suggesting that Ludo's insistence on his students keeping their microphones on may be a good practice more broadly.

> Inspiration within fashion can come from anywhere. From a building, a pattern, or a colour, and often that inspiration is fed back through sound. In this context there's a sense of sound, but without sound, because you can imagine it. If a pin goes into a piece of fabric and it hits a weave, I feel as though I hear the machine making an 'ouch' sound because the pin hasn't gone through as swiftly as it should. Similarly, costumes for historical or period programmes reflect the environment of a time – how different people behaved, how they were expected to behave, and the way the different class systems affected this. When you see them on the screen, you can almost hear these different historical periods coming from the clothing.
>
> Making clothing also makes a lot of sound. The sounds from the sewing machine for example. I have a very old machine that sounds like an aeroplane gearing up when it's switched on. I find these sounds take on a very positive feeling whenever a project is going particularly well. The sound itself doesn't change, but the meaning of it can change a great deal.

I get great enjoyment from sitting at my sewing machine. My daughter has observed that she can always tell that I am in a happy mood whenever she hears that sound of the sewing machine being turned on.

Alyson: UK

Literature concerning fashion and sound, specifically non-musical sound, prioritises the product of clothing itself rather than the process of creation (e.g. Santhanam et al. 2019). As with various other contexts, the primary exceptions to this are investigations into noise and its potential risk to employee health (e.g. Barcelos & Ataide 2014). However, there is absolutely an acknowledgement, if small at present, that the making of clothes is itself a highly performative, kinetic, and even choreographic practice featuring distinctive patterns of sonorous intensity, rhythm, pace, and texture (Nakano 2019). Alyson's comments expand on this relationship even further, suggesting important connections between clothing and physical sound, but also imagined sound and synaesthetic-like experiences in the processes of designing and manufacturing items of clothing.

Sonic warfare

As we discussed briefly at the end of Chapter 5, sound in the military from a phenomenological perspective reveals a noteworthy gap in our current understanding and presents numerous opportunities to explore the everyday human experience of, and relationship with, sound for people serving in military services. Whilst the following commentary I found to be quite powerful, it is also indicative, to me at least, of a great need to explore these sound relationships further.

I would go as far as to say that sound has saved me a lot of times in my life. Growing accustomed to sound contributes to safety. I will instinctively become uncomfortable if something simply doesn't sound 'right'. These effects – at least half of them result from military training. I remember being at a demonstration where a surface to air missile was fired at targets approximately 50–60 meters away, which is very close. The impact of that sound was something that I felt physically, for hours. I also remember being on a firing range and using a particular rifle. The first time I fired this rifle the recoil was so severe, the gun injured my cheek, but this wasn't what stopped me from wanting to fire it again. What stopped me was the sound it produced. It had such an impact that I immediately felt dizzy and nauseas.

Military experience puts a lot of value on listening. I think that listening training is extremely important in any profession that requires self-discipline, survival instinct, quick reactions, interpretations, and decision-making; particularly in an environment where you are placed under stress.

Sound can instil reactions that many people can't even imagine. Not many people understand this, but post-traumatic stress disorder is more often connected with what you hear than what you see. If you are fortunate enough to make it through actual warfare without injury, then you would not physically feel anything traumatic and, often, the environment is such that you will be limited in what you can see. But you are going to hear *everything*, and everything is very loud. This is why sound is such an intense, contributing factor to PTSD. The experience of sound in the military – how it shapes your perceptions and reactions all within an environment in which there is a huge amount of stress and an acute shortage of time – this is very powerful.

Then there's fear. Fear is a factor that stands out even further. It can heighten your awareness and sharpen your perception of sound more than anything else. In certain hands, sound is a weapon. In many military conflicts around the world there are these great, truly terrifying machines that have a lot of fire power to the extent they could annihilate villages with the press of a button. But, in most instances, they don't do that. Because of sound, they don't need to. They just need to fly over an area and that by itself will terrify everyone below. This is exploited, with the idea being to use sound to demonstrate threat and instil enough fear that there's no need to drop a single bomb. Breaking the sound barrier whilst at low altitude is another example. That sound is absolutely a weapon. It's a sound that can shake you. It can disarm you.

Hatim: Pakistan

As Parker succinctly states, "[the] soundscape is a battlefield. The ear is vulnerable, along with the rest of the vibrating body. Listening, in both its physiological and cognitive dimensions, can be commandeered and weaponised; hearing rendered a mechanism of assault, coercion and control." (2019: p.72). Matters of sound as a tool of war have been explored in numerous texts, not least of which is the aptly titled *Sonic Warfare* by Goodman (2010) that explores, amongst many other things, the uses of 'audio assault weaponry' for purposes of crowd control. Other texts have explored the role of sound in military contexts, but largely in terms of the most evocative aspect, warfare. In a 2016 historical review by Meilinger, key features of sound in warfare include: the various uses of musical instruments, from the horn indicative of battle commencement to the trumpets resounding victory; the cumulative human vocalisations and percussions, from war cries and chants to rhythmic beatings of spear or sword upon shield; and various ingenious usages of sound for deception, playing on anticipated interpretations of certain sounds to send false messages to the enemy. In a particularly fascinating example, Meilinger's review also explores the effects of silence, citing reports from pilots in the Second World War who experienced the silent explosions of anti-aircraft fire close to their craft:

danger in the air is an almost 'unbelieved thing' [...]. Anti-aircraft fire explodes near an aircraft and may even shake it, but there is no sound, simply a puff of smoke, like 'black flowers blossoming off in the distance'. This 'unreality' provides a sense of detachment to the aircrew.

(p.81)

Meilinger also reveals a substantial historical context for Hatim's observation of sound instilling fear by way of implying threat. From the 1930s, Ju-87 Stuka aircraft that would activate a mounted siren before commencing its bombing run, to an instance in 1989, in which the US Air Force flew F-4 bombers at low altitude and at supersonic speeds over a base of Philippine rebels: "The message of the sonic booms was not subtle: do not take off again or we will destroy you" (p.81).

In a naval context, the documented role of sound in the military goes back at least a century. In 1920, Hayes published an analysis of hydrophones as an aid to navigating vessels, his report asserting that the technology could effectively determine the direction of submarines and support the exchanging of coded messages. As with several of the environments and contexts discussed above, research on sound and seafaring vessels has largely focussed upon sound experienced within the ship and its effects on comfort and well-being, largely evidencing a strong, negative correlation between the two. In a cruise liner context, for example, Biot and Lorenzo (2008) identify sound (as produced by the mechanical vibrations of the propulsion, ventilation, and air conditioning systems of the ship) as the primary source of disturbance for passengers on board.

I would suggest that sound is a backstop that provides wider situational awareness. Whilst sight will often provide focus on a specific point, your ears are there to capture everything else that is happening in the background. There's an expression in the Royal Navy: 'developing your third ear'. This describes a process of focussing upon up to two tasks at one point in time whilst also maintaining an awareness of information within the environment. This information is external to both of your tasks but may still affect you and what you are doing. As a direct example, if you were in the operations room on the ship, you would have a headset, through which you would listen to both an internal conference call line, but also a 'radio net' line, where you're communicating externally with other ships and aircraft. Traditionally they would play one stream into the left ear and one into the right ear, but then your third ear denotes awareness of people moving around you in the operations room itself. So, colleagues not dialled into the conference call or external radio line, but who still need to convey information to you. People don't naturally have this third ear. It is something most people require training to develop.

I think that different people can have broadly different attitudes to sounds that are new or novel to them. I appreciate my routine and I find familiar sounds comforting. Unfamiliar sounds I find unsettling. This is not consistent across the Navy. I have a friend who talks about how they prefer the vibrancy of new places. I know that, in his position, I would feel unsettled by the unfamiliar sights, smells, and sounds of a new place. When there is an emergency aboard the ship, a fire or a flood say, one of the responses is to switch off the air conditioning. I find that now, irrespective of where I am, if there's a sudden marked decrease in ambient noise, I will become alert, because on a ship, this sudden absence of noise is likely a precursor to an emergency.

Anonymous: UK

That final point above is, sadly, not one that I took the chance to explore further within the interview, but it did make me wonder what possible reason for such a drastic difference in perspective upon known versus unknown sound between crewmembers could be. My instinct suggests this could be a matter of rank and responsibility. My guess is that the officer who feels an acute sense of discomfort from unfamiliar sound, motivating them to attribute meaning and understanding, has greater responsibility for attuning to an environment as a means of detecting and counteracting threat. With regards to the fascinating notion of the 'third ear', this was somewhat challenging to find suitable references for within the academic literature. One such reference, however, goes all the way back to 1986, in a report by Griffin and McBride concerning naval aviation training in the US. Described in the report is a psychomotor task and dichotic listening task that must be performed simultaneously. It was deployed to better identify strong candidates for service training. Participants were required to attend to only the information presented in one ear whilst ignoring the other, all the while using a control stick and foot pedals to maintain a pair of cursors' position upon a visual target displayed upon a screen. More recent research concerning multi-task approaches to auditory training appear more concerned with training machines than humans (e.g. Imoto et al. 2020). However, it is worth noting that, outside of the specific milieu of sound, many of the world's militaries are demonstrating an increasing interest in multi-task training (Talarico et al. 2020), suggesting that exploration of sound in this context is yet another underexplored and potentially fruitful direction of research.

Sound professional by industry

To reiterate, sound professionals, at least based on those I spoke with, are big fans of talking about their work. Throughout the interviews, many of the questions that were intended to be more oriented towards everyday aspects of our relationship with sound ended up relating back one way or another to professional practice. The upshot of this was a diverse and rich set of responses that I found to be both insightful and intimate. However, the sheer amount of information meant there was a decision to make between contextualising these responses with reference to academic literature whilst heavily cutting the responses themselves or documenting more of the conversations whilst limiting the wider context. My decision was to prioritise the contributors' responses. We start with a brief look at the initial responses contributors gave when asked more broadly about how sound featured within their particular industry.

Digital games

The role of the videogame sound designer is to enhance player-experience by creating an appropriate soundscape to accompany the visuals, game-play, and other elements in the game. The sound is there to aid in immersion, enjoyment, and believability. It needs to be appropriate, and it needs to be believable. 'Realistic' can be a problematic term. In videogames, nothing is real, and the setting is often decidedly unrealistic. So, the priority is believability *within* the world being created.

Broadly speaking, sound in videogames can be split into dialogue, music, and sound effects, but it's important for everyone working in specific areas to be considerate of all three. These elements can't exist in isolation. We need to be conscious of how they impact each other. I also think that sound and behaviour are deeply connected. With free-to-play mobile games with micro-transactions for example, these games are quite often centred around making money. Here you will have certain features, graphics and sounds, all designed to be really satisfying in response to the player making a 'desirable' action. These features are designed to trigger that dopamine hit and evoke a sense of reward. The player connects the stimulus to these desirable actions. These satisfying sounds are helping to create this behavioural response. In a lot of ways, sound can be behaviourally manipulative. As a game sound designer, if you look at things in a raw capacity, we're usually trying to manipulate the player, whether it be emotionally or influencing their gaming decisions. This may sound horrible on the surface, but it can be about curating a great experience. It depends on the intentions of the design. If the intention is to affect the player emotionally as part of a positive gaming experience, that's different to encouraging more questionable behaviours such as buying microtransactions or loot boxes.

Sam: Finland, UK

As an audio engineer, I often need to think about how sound can be modified by specific parameters. We can modify individual sound files, but we can also layer sounds in lots of different ways. For example, a car driving past the player. Here, I would need to know the car's velocity and the point at which its position is adjacent to the player, so that rise and fall in pitch that you get through the doppler effect feels accurate. Sound communicates physical data about an object and its properties. When designing that sound for a virtual world, it needs to correctly communicate that physical data. I need to extract that data then present it to the sound designers in a way that they can understand it but also use that data to modify their sound files so that, ultimately, the correct information is presented to the player.

Sound can have a powerful effect on player-perception in a game. I remember the *Half-Life* mod, *Day of Defeat*. It was a competitive WWII team shooter. Each side had different weapons but equivalent weapon classes, one of which was the sub-machine gun. For the Axis, this was the MP40 whilst for the Allies it was the M1 Thompson. Following release, the game's developers kept getting complaints that these two weapons were heavily imbalanced; specifically, that the M1 was heavily over-powered. Statistically, both guns dealt identical levels of damage with near-identical rates of fire and recoil. However, player-data showed that players used the M1 far more frequently than the MP40, but also that the kill-death ratio with the M1 was significantly higher. After investigating, the developers found that it was the sound of the M1 that made the difference. It simply sounded more powerful. This made players *feel* more powerful, and therefore more confident, with tangible results on their performance.

Sound perceived as realistic is a presumed priority in the games industry but, in some cases, the real priority is for the sound to represent other things within the game in the most obvious way for the player. This isn't necessarily the most realistic or physically accurate sound. That said, physical accuracy is still nearly always seen as important, and is of course a top priority in simulation-based games. In such cases, we would start by making the sound as physically accurate as it can possibly be, but then we'll review how the sound functions in context of play. If it's intended meaning isn't obvious, we may then compromise aspects of the realism to enhance clarity and make the sound more useful. In other projects, realism needs to be considered in an unreal context, such as in a science fiction theme. Here, certain things can be written into the lore to circumvent challenges with sustaining realism. For example, sound doesn't travel through space, but we can't have a game set in space without sound, so we create an appropriate fictional narrative. On one project, we wrote that all the ships had sensors that detected events in space and artificially translated them into sound.

Miles: UK

Sound is a tool that we give to the player to support the gameplay. One example of this is location; using sound to guide the player. We may want them to focus on a particular situation or object in the world. Here sound takes on an extra layer of meaning. We often forget that listening to sound is our only truly 360° sense. It gives spatiality to what is otherwise a two-dimensional medium.

Often, what I'm doing with game sound as a designer is trying to breathe life into things that aren't necessarily grounded in reality. The sound has to convince the player that something is real; it has to be anchored to something that the player can identify but, at the same time, evokes a sense of the surreal or hyper-real. Obviously, we don't have laser guns in real life, but the player still needs to hear a sound through which they can identify the function and interaction of the weapon. Creature design is another good example. If we have a monster or alien-type creature that resembles a dog, we may use the sounds of a real dog as a foundation, but then we will build upon that, so it retains some familiarity but otherwise sounds distinctive. There's this duality of linking fantasy and reality around the player.

Anonymous: Mexico, Sweden

In review of these initial comments on digital games development, sound is largely described as complementary or supportive to the visual medium. Meaning is often connected to functions of listening, particularly situational awareness, navigational listening, and confirmation of action. Interestingly, the meaning players can attribute to the qualities of certain sounds in a game is shown to have potentially dramatic consequences in their interpretation of specific aspects of the game. These aspects are not necessarily connected in any actual way to the sound, as Miles illustrates in his description of *Day of Defeat*. The value of more integrated working within creative practice is also underlined, arguing against the siloing of a game's development into specialist strands and towards a greater degree of shared understanding, where individuals may bring specialist expertise to a project but also possess a holistic awareness of the other roles and the key practices and challenges within. Contributors also describe how the interactive nature of digital games brings forth the notion of feedback loops, in which player behaviour drives system feedback which is, in turn, received and interpreted by the player to drive a subsequent response, and so on. Sound is established as a powerful tool in this feedback loop, particularly its potential for evoking subconscious action-reward triggers that have the power to craft a deep and meaningful game experience on the one hand, but also to engrain potentially damaging behavioural patterns on the other.

Arguably the most recurrent issue is that of sound and perceived realism. Here, the definition of 'realistic' appears radically obscured due to the fantastical themes and settings of many game narratives. Hyperreality takes priority over reality; an understandable point of view when escapism is so often the driving motivator for play. Realism often appears to be less about corresponding to a real-world benchmark and more to the players' expectations and understanding of the game. This could therefore be affected by multiple factors, from similar games, they have played in the past, to genre-relevant films or television programmes that, at least for the player, have established a specific connection between sound and meaning. Ultimately, game sound needs to reflect these connections – supporting the player in their actions by delivering clear, unambiguous, and consistent messaging.

Film and theatre

As a film sound designer, I would say that sound has an important function in reinforcing the visuals upon the screen, but also presenting sound that appears off-screen to suggest what could be part of the wider environment. In field recording, you can listen to the soundscape that you're capturing both with and without your headphones on to hear the difference. In most cases, the sound is so different because of the way in which we hear and the way a microphone picks up soundwaves. When you capture a sound, it can feel less natural despite also sounding more detailed, as if something has still been lost. Sound design then becomes about re-capturing what has been lost.

Anonymous: New Zealand, India

Theatre sound is about performance. It's about emotional contact. Sound is an action or event, but it's also a lot more than that. My role considers all aspects of sound; about whether the sound is detailed enough or has enough intelligibility, or about how the sound makes people feel. The sound must match the performance. The loudness must be appropriate. You need to hear everything that is happening upon the stage, and everything must be balanced. This also involves system design, which refers to things like speaker placement. Theatre is about the spatial image; about 'placing' the sound where it should be. If an actor is talking on the stage, the sound needs to be consistent in terms of place. The audience must believe that the actor's voice is coming from the mouth of the actor and not from a speaker. Timing is also very important. There cannot be a delay between the actor speaking and the audience hearing the sound. This is a more naturalistic approach to sound design for the stage and it's not always how things are done, but it is a common approach.

Borneo: UK

A large amount of what I do is making sure that other people's artistic ideals are correctly realised. In musical theatre, a composer writes the song, the orchestrator assigns instruments to perform that song, then it is ultimately my job to ensure that the audience hears that song; that they hear the melody and the details of that orchestration. I would describe this process as implementation and realisation of another person's creative ideals. That said, some fraction of my work with sound allows my personal creative impetus over the top.

One of my primary interests is sound for a cinematic experience, bringing that cinematic feel to the theatre. In the past, theatre sound was expected to be invisible. Microphones were hidden in actor's hairlines. Sound was meant to feel as though it came exclusively from the stage or the orchestra pit. Now we have surround sound technology in our homes, in our cars, even directly in our ears with spatial sound in headphones. I felt that musical theatre sound wasn't changing in line with this. Modern musical theatre should be enhancing the cinematic quality of sound. We should not be apologetic for sound. We shouldn't hide it. Of course, what this also means is that the role of the sound designers and engineers becomes even more of a priority. In shows where most of the sound is coming from the stage without amplification, then the impact of any errors or limitations in the sound that is amplified will be minimised. Once you reach the point that all the sound that reaches the audience is through amplification, now everything needs to be perfectly mixed and equalised because you can no longer hide behind the natural sound coming from the stage.

Gareth: USA, UK

As with digital game sound, commentary on sound in film quickly defines its function in relation to the visual aspect. It reinforces what we see and can extend our awareness of the fictive scene beyond the frame limitations of the screen. Matters of realism are also apparent but there is a more abstract matter of *je ne sais quoi* in which the ideal sound is rarely an unprocessed capture of a physical soundwave, but rather one that has been subsequently crafted. Outside of the composition and production of music, theatre sound is described primarily as a delivery mechanism, one that needs to provide intelligibility above all else. Contributors point to the historical precedent that deeply encourages the 'invisibility' of sound, or at least, of sound technology. Numerous factors, from the architecture of the theatre to the set design, may adversely affect the sound. More often than not, it is the sound design that needs to adapt to suit everything else. Matters of clarity, balance, spatialisation, and localisation of theatre sound will almost always require some form of technological intervention but, just as digital games demand their sound to be perceived as realistic, so must theatre sound be perceived as 'natural'. If sound in a play is to be enhanced with any production process, that process needs to go by unnoticed by the audience – at least that is what we might call the 'traditional view' of theatre sound that clearly, certain players in the industry are keen to progress beyond.

Sound professionals' relationships with sound

In the initial stages of most interviews with sound professionals, there was the consistent impression that they not only thought differently about sound, particularly its meaning, but that their behavioural relationships with sound were also different. This prompted further questioning to directly explore this effect. Several consistent themes emerged that indicated working in a sound profession can have a potent impact on a person's everyday relationship with sound.

Awareness and readiness to capture

The first emergent theme was that of sound evoking greater personal significance for sound professionals. There is a tendency towards always being 'ready to record'. They may notice and attune to the sonic aspect of a particular situation, one in which most people would remain oblivious to how sound is a feature. They may have a markedly greater sense of concern for protecting their hearing, appreciating their ears as a vital means of interacting with and being in the world, both professionally and in the everyday. To lose or damage their hearing would have grave personal consequences. This results in greater auditory sensitivity, particularly to loud, sharp, or sudden sounds. It also affects overt behaviour, such as greater likelihood of wearing earplugs in everyday situations or avoiding environments that may contain damaging levels of sound.

> I often carry a microphone with me and record different unusual textures. You never know when you're going to need it, and it's often exactly what you need at a certain moment.
>
> Mark: Russian Federation

> I'm a lot more hyper aware of the sound around me when going through my daily life. But that's also why it's good to always have a pocket recorder with you, since you don't know when an interesting sound might pop up.
>
> Rasmus: Denmark, Sweden

> There are moments when I find myself analysing what I hear more carefully but it's also important just let the emotions take you to other dimensions and go through the experience analytically later.
>
> Heikki: Finland

> I'm always looking for interesting sound, which is to me sound that has an unusual acoustic profile, such as a particular reverb, flange, or distortion quality. Reverberation of spaces also comes to mind. I feel working as a sound designer makes me very conscious of the different reflective properties of interior spaces. If I'm outdoors and I hear an interesting sound, I will record it if I can.
>
> I find that there are sounds outside that really bother me, but looking around at other people, I notice that no one else seems to be affected. I'm always thinking about sounds as potential professional hazards. My ears are effectively my tools so I'm very conscious of things that could

damage my hearing. If I'm at a movie theatre and the audio is setup too loud, say, over 85 decibels, then it genuinely hurts my ears. Yet, the rest of the audience is unphased by this.

Arturo: Costa Rica

In terms of building or renovation work around my house, I am far more concerned with the acoustic aspects of the result, whereas most people would be more interested in visual aesthetics or structural integrity. Instead, I'll be considering how much insulation there is in the wall cavities, what effect different rugs or curtains will have, how loud shutting a door will be. These definitely result from my professional background, which has encouraged me towards being an amateur acoustician. I think this experience encourages me to consider the isolation of sound but also more qualitative aspects of acoustics. Ways of embellishing sound and making it prettier or improving intelligibility is something that I think also bleeds into my everyday life. This does translate into frustration sometimes, as I find that I'm consciously aware of the wash of reverb around many everyday sounds. It feels as though I've unlearned a piece of human evolution in my attention being distributed like this.

Brecht: Belgium

I had no idea working is sound design would make sound take over my daily life in the way that it has. I find that my work makes me more sensitive to sound in my everyday life. The small details of noise around me. My neighbours. Air conditioners running. These sounds I find do bother me more than other people I know. Outside of my house, at any given moment, I can be distracted by a sound. It just needs to be something that I find unusual. I think about these *new sounds*, and I immediately think about the possible process I might use to recreate them.

Can: Turkey, China

I find that it's difficult to sit down and just watch a film. I'm constantly analysing the sound, how it's used to support the storytelling, how it may have been produced. I will consider the qualities and possible meanings of general sound out in the world. I'm always searching for rhythmic patterns in sound, everywhere I go. If I'm at a café or restaurant where there is music playing, I might find myself comparing the rhythm of the music to the footsteps of the people passing outside to see how closely they match. I think that, as humans, we're inclined to perceive rhythm and tempo in everything. Before we are born, one of the most significant elements of our sensory environment is the rhythm of our heartbeat. Rhythm can be a way of subconscious social bonding. I think of places such as a hospital waiting room where people keep to themselves and are unlikely to start conversations, but if there is music playing, many will be subtly moving their feet in time with the rhythm. It's contagious. People

will see others around them doing this movement and without thinking will start doing it themselves. It's quite a beautiful thing to watch
happening – to think that we are all connected by the shared experience
in listening and moving to the music.

Andreina: Venezuela

Many contributors reported this feeling of responsibility towards always being prepared to capture a sound or a soundscape, particularly one that they
found to be unusual in some way or of possible application to a current project. Often, but not always, in service of this preparedness to record sounds,
several responses also document some form of enhanced acuity towards
sound. In some descriptions, this is largely attentional, pulling the listener towards sound objects and events that many people would instinctively ignore.
In others, it emphasises greater analytical comprehension of the soundscape,
their comments highly reminiscent of Tuuri and Eerola's (2012) reflective
listening modes, namely analytical (where the listener examines the objective
acoustic features and qualities of a sound) and critical (in which the appropriateness of those acoustic features in context of the function, setting, or wider
aesthetic is judged). One particular comment raises the intriguing notion
that working with sound can not only skew a listener's instinctive reactions
to sound towards reflective listening modes but that the listener may respond
to this effect by turning their conscious attention towards the reflexive and
denotative listening modes that most of us don't have any real awareness of,
such as observing the group-entrainment effect of music on foot-tapping in a
waiting room. Another comment takes things a step even further, suggesting
that critical listening in particular can manifest as a heightened awareness of
acoustic ecology, in which the listener not only builds a rich understanding
of the sound in context, but also of the numerous non-sound components
within the soundscape, how they are affected by the sound and, in turn, how
their resultant behaviours complete the loop, contributing to the very same
soundscape that first affected them. The cycle continues.

Impact on everyday enjoyment

In the previous section, we observed how many of our contributors tended
to be sensitive and analytical towards sound. This has a distinct knock-on
effect here, as further comments identify ways in which this instinctive drive
towards attending to, analysing, and critiquing sound can impact engagement
with media such as film, television, stage shows, and digital games. In some
instances, these effects can be less than desirable.

As a sound designer, I can't watch movies the way I used to anymore.
Whenever I do, I think about how to mix them, how the sound effects
were made, what kind of mics they used, and so on.

Anonymous: Indonesia

I've worked with colleagues who cannot abide music on cheap phone speakers, but I've never had a problem with that. There's no high or low end and the signal is compressed to hell, but it doesn't bother me. That's not to say I don't enjoy listening to music on my high-end home setup, I absolutely enjoy good quality sound, but it's not an essential part of my daily listening life. When I go to a live show however, particularly a musical theatre production, I will really struggle with poor quality sound. It feels really frustrating. Many cinemas show films with different audio technologies, and I do admit, I will always go for the showings presented in IMAX and Dolby Pro.

Gareth: USA, UK

I find that I don't listen to music very often because I need my ears for work throughout the day. I'm often a little burnt out from listening to sound all day. I don't really listen to much music after work, except when I'm exercising. My partner uses podcasts and audio books to fall asleep, but I really can't do that. I think one effect of my profession is that if I can hear sound, I feel compelled to focus my attention on it, so I can't use it as a passive background. If I'm trying to sleep, or if I'm working on a paper or a lecture, I'll pretty much need complete silence, otherwise the sound is too distracting.

Michael: USA

It's difficult for me to disassociate sound in my work and in the everyday. If I'm watching a movie for instance, I find myself analysing the sound. I will consider how the sound was probably made and think about how I would create it myself. In recent years, I have been really interested in interactive audio and this makes me think instinctively of the elements of soundscapes; how individual sounds could be captured and combined in an interactive context. My professional focus is more on sounds generated through artificial processes, and I have found that this instinct to analyse sounds is much more pronounced when listening to man-made sounds. I find that I enjoy natural sounds more as I can listen more passively and don't feel the same need to listen analytically.

Anonymous: New Zealand

I tend to pay more attention to the sound design while watching movies and television. Sometimes this makes me miss a line or two of dialogue and I need to rewind a few seconds to make sure I don't miss a plot point.

Marc: Mexico, China

There are certain situations where I hate listening, then others where I feel I must have sound. When I take a shower in the mornings, I need music. When I cook, or do other everyday tasks at home, I need to have music on in the background. But after a long day of working with sound, particularly when I am going to bed, I don't want to listen to anything. I live beside a small creek which produces that trickling sound of running water. Everyone I know tells me 'It must be so incredible hearing that relaxing sound every night'. I always tell them the same thing – I hate it. The sound is similar to white noise, and I spend all day working with sounds that often share that characteristic. So, at home, I just want to shut it out.

Daniel: Costa Rica

Throughout these comments the notion of listener fatigue is also raised on several occasions. It suggests that it is possible to reach a saturation point in our cognitive and affective relationship with sound where we feel the need for some (sonic) space – some time apart and some indulgence in silence, so that absence can make the heart grow fonder. Unfortunately, it appears as though an intensive working relationship with sound can have long-term implications when it comes to consuming music and sound recreationally. It appears the old adage of not indulging in office romances extends to sound. Become too involved and the distinctions between work and homelife blur.

Fortunately, not all these effects of professional practice upon everyday sound experience are negative. Several contributors explain how their work with sound enhances their appreciation for it which, in turn, can drive them to explore sound much more and ultimately discover and experience a wider and more diverse array of media content. There were also several occasions on which I couldn't resist expressly raising the question of snobbery, specifically whether working in sound professionally meant that a person's tolerance for low-quality audio would be diminished. My hypothesis is that one couldn't be a sound professional and not also be an audiophile. What I found in asking this question, however, was that whilst many acknowledged their enjoyment of exceptionally high-end sound technology, this did not dictate any kind of aversion to more average means of audition. I'd also like to point out that not a single contributor seemed to take any offence to my suggestion that they could be a sound elitist. Perhaps they were used to people making such assertions.

Changes in perspective

In certain instances, respondents took the discussion all the way back to our very first question on sound, namely the ways in which a person establishes meaning in sound, how they definite it, and what it means to them. Many observed that working professionally with sound had changed what sound meant to them at a fundamental level. This occurred in their reflections but, in some cases, contributors made realisations about their relationships with sound during the interview itself.

I would say that my interpretation of sound before I became a sound designer was very bland and uninteresting. I wasn't particularly focussed on its possibilities. As a designer, I can experience these great stories unfolding all around me through sound. There's also the effect of knowing the common audio libraries when watching films and television. Many productions use the same sound recordings repeatedly. So, the sounds of a creaking door or a dog barking become these sonic cliches that you can't help but recognise.

Arturo: Costa Rica

Previously, more as a musician, I would consider sound more in the abstract. From the time I began working in music and sound production it is much more about the physics of sound – soundwaves, physical properties, a vibrating medium, and so on. So, right now, I would think of sound mostly as an object. I find that producing sound and music is not so much an emotional process as much as it is a part of daily life in which I am mainly producing whatever it is that a client needs.

Anonymous: Serbia

I think that sound is something that most people take for granted. But when you start working with sound, such as for film or theatre, you become more aware of the difference that good sound can make. Then, as you understand more, you become even more interested and aware of sound around you, which helps you develop further.

Angélica: Spain

I think both working with sound and studying it for my dissertation have changed my everyday experience of sound. Researching sound, it of course becomes a real focus of your attention. It makes me realise the extent to which sound is everywhere. Working in sound design, then watching movies, I start to recognise the sound libraries being used. I also find that when I hear sound in film that I think is good, my mind starts wandering away from the film and I start thinking about the design process – about how the designers made these sounds. Working in sound also makes me think about the purpose of sound in films and video games a lot more than when I just watched films or played games. You appreciate how sound is so necessary, how it identifies things within a scene or provides instruction to a player. I think this affects how I listen in everyday life. I focus on what sounds mean and the information that they can tell me a lot more than I used to.

Anonymous: New Zealand

For some, knowledge and experience of sound design meant that the meaning they attributed to sound as a consumer had irreversibly changed. Maintaining suspension of disbelief in a film for example, and accepting the intended meaning of sound, understandably becomes a lot more difficult when every sound you hear can be immediately identified in terms of which sample library it came from. Indeed, once you identify the Wilhelm[1] or Howie[2] screams as cliché, you cannot unlearn those associations.

The idea of professional practice in sound drawing listeners instinctively towards more reflective modes of listening is extended to a connection between analytical listening and interpreting sound as an object of analysis. Defining sound primarily in functional terms is also a recurring feature of the responses, including navigational (sound helps us find and reach target points in space), empathetic (sound helps us infer the emotions of another and adjust our behaviour accordingly), and spatiotemporal (sound gives us rich descriptions of an environment, see Grimshaw's [2007] choraplast, topoplast, aionoplast, and chronoplast – discussed in Chapter 1). Whilst contributors would seldom quote any theoretical models discussed at the start of this book, many inadvertently described them in accurate detail, demonstrating a great practical understanding of sound function, if not a declarative one.

Perceptual effects

The final theme that emerged on the question of how working with sound could affect our everyday relationship with it concerned two possible changes to perception. The first overlaps with comments on instinctive analytical and critical listening.

> I would say that I hear everything in music notes. I have absolute pitch, and so I will hear non-musical sounds as notes. If a sound isn't quite at the correct frequency to be a recognised musical note, then I will hear it as a 'dirty' note. This isn't something that I explicitly learned how to do, it developed naturally as a child, so I think it's more about my early musical experience. In my professional life as a composer, I developed the ability to isolate any source of sound by mentally shutting out everything else. I use observable differences in frequency to do this. I have found that, in everyday listening, I can also concentrate to close my mind to sound.
>
> Tim: Russian Federation

There are certainly resonances of synaesthesia in the comment above. From the responses received, there was no clear indication of whether such effects have a positive or negative impact on the listener's general wellbeing. Of course, it certainly cannot be anything other than an asset, to a sound engineer especially, to have this ability. That said, whilst being a gift in one context, it is unclear the extent to which such a perceptual effect may be of benefit or detriment to daily life. The second perceptual effect was one that I found deeply compelling because it was all too familiar – the sense of having a reduced ability to focus in upon a singular stream of auditory content within a sonically busy environment.

> I sometimes find it very difficult to focus my listening on what an individual person in saying to me. In a crowded environment like a restaurant or bar, I feel as though my brain is so wired to listening to the

soundscape that I've lost the human ability to focus in on one conversation. Nature gave us the ability to selectively hear and to ignore or block out anything that doesn't have immediate relevance to the situation, but I have real trouble with that.

Daniel: Costa Rica

I also find that the cocktail effect doesn't work particularly well for me. When there are a lot of different voices in the room, it's difficult not to move your attention between them and instead focus just on one voice.

Michael: USA

Now, yes, I did say a mere handful of pages prior that I wasn't going to contextualise the commentaries of this chapter within academic literature, but please indulge me in this exception. In 2006, an edited work by Augoyard and Torgue entitled *Sonic experience: a guide to everyday sounds* offered readers a single point of reference to a huge array of what the authors refer to as 'sonic effects'. These effects explain a great number of perceptual phenomena associated with the experience of sound and are extremely relatable to both sound professionals and non-sound professionals alike. The effects are grouped into six classes, some helping us to describe sound objects, events, and environments, others giving us terminology to help us explain and understand our perceptual experience of sound:

- *Elementary*: basic acoustic effects of an environment upon its soundscape (e.g. echo, delay, filtration)
- *Compositional*: commonly associated with music but that can also describe features and qualities of non-musical sound (e.g. accelerando, crescendo, the Tartini illusion[3])
- *Electroacoustic*: technical effects within sound signal processing (EQ, flange, fuzz, etc.)
- *Semantic*: matters of contextualisation and attribution or de-attribution of meaning to a sound
- *Psychomotor*: various phenomena in which sound affects physiological state or behavioural action (a good example of this is the Lombard effect (see Zollinger & Brumm 2011) which describes an instinctive tendency to increase our vocal loudness in response to increasing background noise)
- *Mnemo-perceptive*: phenomena that arise from the interplay between the soundscape and core elements of the mind, including imagination, attention, and memory. This includes the 'cocktail party effect' — the ability to focus the attention upon one single auditory stream of information, typically within a noisy environment and for the purposes of communication and social interaction

For anyone interested in exploring psychoacoustic phenomena in relation to everyday listening, I strongly recommend Augoyard and Torgue's work on the subject as there are so many fascinating effects and there simply isn't the scope for me to discuss even a decent fraction within this chapter. Now let's return to our contributors' belief that, in working with sound, they have lost, or at least diminished their ability to utilise the cocktail party effect. They refer to those earlier comments we observed regarding enhanced soundscape awareness and sensitivity to sound sources, suggesting that the inversion of what could otherwise be seen as advantageous to situational awareness and their professional practice, can be disadvantageous when the requirement is to focus on a single stream of auditory information and to perceptually to shut out any extraneous sound. At the eponymous cocktail party, we don't mean to be rude, and we're not (necessarily) bored in what you're saying to us, but within our own perceptual experience, every little sonic thing within the space is vying aggressively for our attention.

Cultural factors, expectations, and creative response

The relationship between sound and culture with regards to professional practices in sound is nothing if not complex. Many different factors and variables, connections, and feedback loops collectively form a dynamic, emergent profile. This profile defines the sound professional, their perspectives on the work, and their approaches and attitudes towards their practice. Although one of the more extensive sections of the book, there is no pretence here that we shall do more than scratch the surface, but many of the responses below I found to be some of the most intriguing of all the insights I collected.

Cultural effects on practice

When talking about cultural effects on the practice of working with sound – how culture may change preferences towards certain approaches and strategies – contributors described a range of alternate factors. One frequently mentioned effect was that of specific events or moments that contributors felt to be influential elements within their personal histories, but there was also a common sentiment that nationality and multiculturality also played a part.

> I think that my personal quest for discovering things that are novel and interesting – new, unexpected tones, dynamics, and textures – this goes back to my upbringing and the more unusual music that I was exposed to as a child. I'll never forget when my brother brought home a 12-inch vinyl, *Let's Get Brutal* by Nitro Deluxe. At the time it was like nothing I'd ever heard before. That love of surprise I had when listening to it was certainly something that fed into my professional practice.
>
> Stafford: UK

> I think my work and my perspectives are more heavily influenced by the culture of being a sound designer, than of being Venezuelan. I've worked

on projects with sound designers from many different countries. I find we all have similar relationships with sound *because* we are sound designers. I would agree though, that being from Venezuela has an effect. I can be more emotional when it comes to describing sound because Venezuela is a very emotionally expressive country. Venezuelans are typically very warm, socially open, and welcoming. That could have some influence on my perspectives on sound, particularly on how I talk about sound. I think it also affects my analysis of sound. Whilst I will consider technical aspects, I do find that I prioritise the emotional aspects of a sound, how it can change my emotional state, tell me a story, or change me physically.

Andreina: Venezuela

I remember, whilst working with several Arabic musicians, I learned interesting techniques that I would have never used before. Different cultures think differently about sound. They may choose very different sounds for the same purpose. Sounds that are too high frequency for the European market could be fine in Asia, whilst the same song produced in Mexico would sound very different if mixed by a German engineer.

Anonymous: Germany

When we talk about culture, it is very individual. My collective experience of the media I've consumed, the people I've spoken to, the different cultures I've experienced living in as well as the virtual gaming cultures I've been a part of – all these things will affect how I design sound. Then there's the cultural effect at the team level. I believe if everyone on a sound design team is from broadly the same culture, that will have a difference on the practice and the output when compared to a team that is more diverse and multicultural.

Sam: Finland, UK

My own set of tools and tricks come from previous teachers and colleagues, and I think making a greater number of bonds with others from different backgrounds is key to developing your professional practice as a sound designer. I think that our practice can be enclosed in a national sense, where the circle of people teaching and sharing sound design can be limited to a specific place. When colleagues come to work with you from outside that circle it can have a significant impact. One of my foley teachers who came to Costa Rica from Argentina told me that she had been travelling to several places specifically to learn new sound design techniques. She wanted to go to India because she had been told that the way they did foley was very different and she wanted to expand her experience.

Arturo: Costa Rica

Here, repeated exposure to new musical styles and genres is an interesting effect, described as being strange and novel to the listener and consequently

generating not only a wider appreciation for music but also a more foundational appetite for experiencing new things. Another thought that warrants
further consideration is that of working with sound being a potent influence
upon, well, working with sound. In discussing the nature of the industries
within which they worked, many contributors described what they felt to be
clear cultural differences between certain places. Adjacent to this point is the
intriguing proposition that working with other sound professionals from a
wider array of cultural backgrounds may be a powerful means of developing
as a professional, both in terms of knowledge and skillset. Contributors often
further suggested that their experiences of working across multiple countries,
or collaborating with other professionals from other countries, often presented substantial opportunity to expand their sound design 'bag of tricks',
with new techniques and fresh perspectives.

Universality and concerns over detracting from the art

As the discussions on culture and professional practice continued, many
contributors turned away from the practice itself and towards sound as the
product. The sonic fruits of their labour so to speak. This raised numerous
questions on the extent to which features or qualities of these 'sound products' are ever required to reflect preferences, expectations, or even demands
that derive from cultural factors. Within this topic, one notable and repeating
perspective was that 'good sound' was either largely determined by universal
human factors, or that good sound *should* be determined by universal human
factors.

> It's our job as sound designers to create a general response to sound for
> the whole audience. The sound needs to communicate the same narra-
> tive and the same emotion to everyone in that audience. We can't leave
> anyone out of the story.
>
> Daniel: Costa Rica

> I think that there are some very human effects when it comes to per-
> ception of sound. For example, when I was working on rock and roll
> performances, we would always put the walk-in music through a re-
> verb unit and as the audience were walking to their seats, the music
> they heard would really be echoing. This would set an expectation for
> the audience. They wouldn't consciously realise it, but it would set this
> precedent that they were walking into a very reverberant space. Then,
> when you start the show, the sound is no longer going through a reverb
> unit, and everything now sounds so much cleaner and tighter. Without
> realising the mechanism that we're using, the audience experiences a
> difference that is really very powerful. Another trick is to place micro-
> phones around the audience which we activate during applause, to cap-
> ture and amplify the sound. This gives the impression that the applause

is greater which, in turn encourages the audience to actually clap louder and for longer. These microphones can also be used during the audience walk-in, in combination with the reverb effect I mentioned. Now, if somebody coughs it seemingly bounces around the theatre. Then all of this is switched off as the performance starts and everything becomes clear, tight, and clean.

Gareth: USA, UK

With sound design you need to tell a story, convey a feeling, or explain a concept. But you must do it in a way that's universally understood. For that we can use different techniques that are almost universal. Things like the Super Mario coin-pickup sound have become so engrained in society that whenever you want to convey something that indicates achievement, it's a good approach to make the sound go 'baa-ding!' because so many people have that association.

Rasmus: Denmark, Sweden

In terms of culture influencing the sound for marketing a game to a specific demographic, I can say that the company I work for actually seeks the opposite of this. I would say that the intention, as best as I understand it, is to avoid targeting a demographic and to create a more all-encompassing player experience.

One thing I can mention, that's almost a reverse of culture influencing sound, is the use of sound to help represent different fictional cultures within a game. Science fiction games will often be set in the distant future and the lore will typically have several very different, alien cultures within the fictional universe. Different characteristics will be assigned to each culture, and this will mean that specific sounds are also assigned to them. Here, the writers will produce a very detailed synopsis of each race, down to the very fine detail, then the sound design team try to produce the sounds that best characterise each of the details to really embody the races through sound.

Miles: UK

Here, we have some wonderful tricks with universal application. I challenge you now to *not* notice the reverberant incidental music being played through the loudspeakers in the atrium and corridors of your local theatre, as you make your way to the auditorium. There is also a clear sentiment in the intention for sound to create an identity, either alone or as part of a larger multimedia work, rather than have identity thrust upon it from some external source. In particular, the desire to protect the integrity of the story is an important expression to observe and it suggests that should culturally related expectations grow, and should this lead to restrictions on the ability of a creative community to tell the story that they wish to tell, there will absolutely be a great deal of resistance.

Different standards and expectations

There are many instances of artistic works being adapted for different audiences. In digital games, Grimsley (2020) cites the dramatic visual redesign of the title character in *Crash Bandicoot* and the removal of multiple violent and sexually explicit events from *Grand Theft Auto V*, to suit the (presumed) more anime-orientated aesthetic preferences of Japanese consumers and the (definite) requirements of Japanese censors respectively. Within film, and emphasising something of a Japan-America dichotomic trend which we'll return to, Jiuliani (2021) points to Pixar Studios' swapping out a child's disgust at broccoli for green peppers in *Inside Out* and the inclusion of an introductory preamble by Stephen Spielberg to contextualise the biopic *Lincoln*, both to meet what were felt to be cultural needs of Japanese audiences. Adaptation to a global market is an increasingly popular subject and I was highly curious to ask contributors if they had ever adapted their sound content based on cultural standards or expectations.

> Nowadays, because of streaming platforms like YouTube, there is a lot of amateur content being created that does not really consider sound. You can hear sound on these videos with really horrible reverberation for example, but you can see these videos still have many views which I think means a lot of people are getting very used to 'bad' sound.
>
> Angélica: Spain

> Certain markets that include France and Spain require their languages to be louder within the mix to meet standards for intelligibility. The English language by comparison doesn't typically require that added volume. I've encountered this issue myself in terms of music mixing, but it can also apply to sound for visual media. Of course, exploring and validating these sorts of things comprehensively across the world would be very challenging, but these are very interesting questions to ask of sound design. I don't know to what extent production companies may be directly considering these issues, but I have noticed that there is a definite range of accessibility in terms of English television programmes to viewers who are not English native speakers. I think factors that influence this include the extent to which regional accents and dialects are used, but also the tempo of the actors' speech, and the speech-exclusivity within the mix. I find that producers targeting a wide international audience will favour Received Pronunciation and will make sure that other sound effects have minimal interference against the dialogue.
>
> Brecht: Belgium

> Iceland is a very small country and when you produce television, pretty much everyone is going to watch it, so you need to try and please everyone. The requirements are simple: clear dialogue and music.

No atmosphere, no foley, no crafted sound is needed. For a sound designer there's not much space for inventive, creative mixes. This is something that I've been fighting against for quite a while. I've been moving away from that emphasis on clarity over creativity, because now with Netflix and such services, we are reaching larger and broader audiences. This has begun to widen requirements and has been something of a relief to a lot of sound designers.

Nicolas: Iceland, France

I would say that every genre has very different expectations that you need to match. Different cultures can be more interested in certain genres but, even within a single, or highly similar genre, there is still some cultural variation in terms of sound. For example, a Hollywood-produced musical will sound different to a Bollywood film. Mobile games are typically distributed online, worldwide, in a variety of different languages. Here, a more global perspective is essential. How will the player perceive the English dialogue lines when the game will also be showing translated subtitle text? Are certain sounds in one country perceived differently in another? I recently mentored a sound designer, based in Yemen. He sent over a commercial that featured a camera panning over a mosque. I commented that the sound during the shot felt empty and flat, particularly as there was no sound attributed to the mosque. I suggested that he add a sound clip of the Adhan for the shot. The response from the designer was that they couldn't do this because it would be culturally offensive.

It's worth mentioning that, with the prevalence of various spatial sound technologies such as Dolby Atmos and multi-speaker systems, sound designers are mixing differently based on the required output format. Depending on how many speakers you have and their configuration, that can actually change how much audio is in the film. If you only have monophonic sound, you will have to heavily prioritise certain sounds during the design and mixing process. In this instance, when prioritising sound during mixing, dialogue is king, and music is queen. This means that there simply isn't the room for a rich soundscape. Lots of sounds will be cut for the mono mix. With a more complex, multi-speaker format, there is now much more room to spread out numerous sound sources, allowing a much more complex soundscape to be presented. This means that some high-end versions of a film will feature sounds that simply don't appear or aren't audible in the downgraded versions. Now, because different countries have very different audio technologies in their movie theatres, this means that certain countries can experience sound in the same film very differently to others.

Michael: USA

Whilst being easily understood is more important in certain countries, I find that sound design in broadcasting, for example in UK or US television, allows for a greater dynamic range. Sound is heavier, thicker, with more impact and better balance between speech, music and all other sound. In Germany, although there is a movement for a more modern view on dynamic range, we are more bound by these standards for intelligibility. So much film and television consumed in Germany is overdubbed in German, so our audiences are used to hearing these speaker booth-produced voices. They have become the norm. It sets an expectation for sound in broadcast production that audio engineers need to match. In the German film industry, there is currently a discussion being had over dynamic range. With big Hollywood productions in particular, the filmmakers are increasingly demanding that their overdubbed versions have the same dynamic range as the original film. This means the speaker booth voices may be starting to get phased out

Iñaki: Germany

I don't think that there is much established understanding on cultural background influencing expectations for sound quality in theatre, but there could be some anecdotal instances. For example, you're presenting reggae music to a culture of people who were brought up listening to that genre, there will be a greater expectation for more bass in the mix. They will interpret 'good sound' as being lots of bass. If you were delivering a musical with a reggae number in say, Jamaica, maybe it would be worth considering how their expectations for the sound could be different.

Gareth: USA, UK

Demographic is an important consideration in games, with sound being a relevant part of that. A great example would be gore sounds. We can have a game that is aiming for a 'teen' rating, and so the design doesn't include any visual blood or gore even though it's depicting a violent interaction. Here, the sound design team needs to be consistent with this and be careful not to include anything like a gut-splatter. If you were creating a *Doom*-like shooter but for a younger audience, you may be able to include, for instance, the chain-saw sound, but you may take out some of the low-end to make it softer and less visceral. Because there is an overriding importance to always stay true to the visuals, any way in which those visuals are presented to target a particular demographic needs to be reflected in the sound. We can't have a cartoonish moment in the game but with blood and gut sounds everywhere, that would destroy the immersion.

Anonymous: Mexico, Sweden

There are some differences in what quality means between certain places. For me, the American standards of sound are my basis for what I would describe as quality. But I think that quality is actually lower than ten

years ago. We have YouTube, Vimeo, TikTok, and other video streaming platforms where too many uploaders don't consider audio quality,
so people are constantly exposed to, and will expect and tolerate, poor
sound. Many professionally produced television shows also accept a poor
level of sound quality, and this standard is continuing to get worse. There
are some shows where the audio has no element of production or processing at all, it simply gets captured by a lapel microphone and that
recording is added to the visuals. That's the complete pipeline. This is
something I see quite consistently in markets around the world, not just
in a few specific locations.

Tim: Russian Federation

Of the various perspectives presented above, technological developments
definitely raise some highly pressing matters. What I found genuinely surprising was the observation that the standards and expectations of different
audiences also include an important distinction based on their preferred
media-consumption platform. This emphasises age and generational identity as key cultural lines. Internet connectivity speeds and mobile technologies have empowered small or even zero-budget content creators to
reach worldwide audiences and caused a seismic and fundamental shift in
the way huge swathes of the population (myself included) consume media.
As observed by our contributors, the implications for audience expectation
could be immense. It is certainly true that a great amount of online media
content features low-fidelity sound, though this is not a constant, and there
is growing evidence that content creators are becoming increasingly aware
of the need for improving the quality of their audio, whether publishing
pre-recorded material or even when live-streaming.[4] As also observed by
our contributors, the content distributed by these online media platforms
has global reach, meaning consumers can be exposed to content from potentially anywhere in the connected world. As a result, great numbers of
people are now experiencing sound production standards and aesthetics
outside of their norm. A norm that may have previously been rather paro-
chial and nationally bound.

Whilst more than a few contributors stated that adapting the sound product itself with regards to cultural factors was not something they had any conscious awareness of doing, some responses were quite emphatic about doing
so but equally emphatic about the *need* to do so. This was evident across film,
television, and digital games, whilst less prominent in theatre, and with certain variations in what the priority adaptations were between industries. Scenarios could include those where the ambition is to subtly reflect the nuances
of certain cultures with minor ornamentations as a strategy for expanding
into a new international market, acknowledging that different cultures can
have different expectations and preferences. It could, however, be something
more critical, such as deploying cultural sensitivity to avoid causing offence
and potentially alienating an audience.

The return of the East–West dichotomy

It is an arguably safe assumption that different cultures across the world have some discernible specifics in their expectations, preferences, and even requirements for sound products, be that in the context of cinema, television, stage, or digital games. Precisely what those differences are, however, is less certain, with even some of our sound professional contributors expressly saying that there are differences, but they simply don't know what those differences are. This of course presents us with a real opportunity for further exploration, potentially moving towards a worldwide framework of cultural expectations for sound, but this is well beyond the scope of this book. That said, several responses from contributors did begin to unpick some of these possible differences, particularly in broader terms of contrast between Eastern and Western societies.

> I haven't experienced this myself, but I've been told that some eastern audiences won't clap during a performance, only at the end. This doesn't in any way mean that they're not enjoying it at the time, and in their applause at the end they'll be extremely expressive, but there's this sense of being respectful and not interrupting the performance. This is obviously a big cultural contrast with many western audiences where you have them clapping throughout, particularly at the end of a song or in response to an especially funny moment.
>
> Borneo: UK

> In the game company where I work, we usually have two different versions of the same game, one for the local market and one for abroad. Although it hasn't happened to me yet, I'm aware some of my colleagues need to create two different sets of sound effects, one for the Chinese market, where they dial down the gore sounds, and another, more visceral set for the overseas version.
>
> Marc: Mexico, China

> From my experience, there's no real way for me to say that every country has its own style or cultural touch on sound, but there definitely are differences that you can observe between wider regions across the world. For example, sound design in Western-produced games certainly has differences compared with Eastern games. The Japanese game industry especially has specific sound design features that feel unique to that culture. Then of course there is the Hollywood style of blockbuster films and games in the North American entertainment industries. Between these two cultures I think that the main way of describing the difference is that the American sound is more about exaggerated physical realism, whilst the Japanese sound is more about distinction and character. I think of the sound of Mario jumping as an example. It's a sound that, despite

being synthetic, is immediately recognisable as 'the Mario jump', because of its unique design and character. Footstep sounds are also a good way of explaining the difference. American design will require you to create many variations of a footstep sound, even if it is for the same character on the same surface, so that it doesn't sound repetitive. Japanese design has the opposite approach. They intentionally make fewer variations, maybe no variations at all, precisely so that you hear the repetition – you experience the same sound with more frequency, reinforcing your attachment of meaning to that sound.

Can: Turkey, China

I do think that sound design can be influenced by culture. Distinctions have been drawn between the Eastern and Western audience markets, with some game companies being known to adapt their product for a particular demographic. There will be different perspectives on sound design, on the mix of a soundscape, and priorities on what the player should hear and when. Designers are often influenced by other cultures. For example, there are a lot of fans of Eastern approaches in the West and vice-versa, but these are stylistic choices a lot of the time. I wouldn't say that there are any absolute rules, and these are largely anecdotal observations, but I would expect looking deeper into this, you would find trends such as hyperbolic or hyper-real sound favoured more in certain parts of the world, with authenticity the priority in others.

When you consider sound design for games in regions such as Japan, China, or Korea, you may find that repetition, order, and relationships with sounds are preferred. In a lot of Japanese role-playing or fighting games for example, there's a perceived value in using repetition to reinforce the association between a sound and a particular action. There is a sense of order there. By comparison, Western markets often feel that repetition is something to avoid, something that can break the immersion. It's worth noting that the Western approach to sound design has been dominant for a long time, thanks largely to the Hollywood movie industry. But now that international media is more accessible and having an increasingly powerful impact globally, we see more and more influences appearing from different cultures.

Sam: Finland, UK

In Western cinema, the Hollywood sound is always over the top. It's always as much sound as you can add. Anything that moves on screen needs a sound, but sound often needs to reflect what's happening off-screen too. I review sound design by other designers from around the world, and I find that I'll often feedback that they're missing sounds; that the soundscape is lacking the required detail. If you compare the Hollywood sound to movies from places such as Japan, or South Korea, the sound design is very different. In these cases, it's about identifying the singular, most important sound to add for that moment.

Michael: USA

You may find more reverb in film sound within Indian Cinema compared with other places, but that is itself changing over time as the industry develops. Modern Hindi cinema is less likely to feature that same amount of reverb as it once did.

Anonymous: New Zealand, India

The above comments may not represent a comprehensive review of cultural differences in designed sound between the broad dividing line of Eastern and Western societies, but in identifying a good number of points, they certainly reinforce the assertion that 'good sound' is not the same everywhere. Productions wishing to find success in certain markets would be well-advised to proceed on the assumption that different audiences will indeed have unique tastes and priorities for the sound they consume. This appears to be the case not only in one industry but across all those that we have focussed on throughout this chapter, such as the observation of Eastern preference towards repetition as a means of conceptual reinforcement appearing both in discussions on film and digital games. Intelligibility may be important, but it is not equally important everywhere. The dominant force of the Hollywood sound will not have the same impact in certain parts of the world, but sound designers also cannot rest on the assumption that what works today will work tomorrow. The 'sands of preference and expectation' are nothing if not shifting.

A few further examples

As we approach the end of this chapter, I wanted to enclose a few comments that stubbornly refused to fit neatly into the above structure, but that I simply couldn't allow to be edited out of the book.

Back when I was working on soap operas, there was this scene where a character went to the toilet and there wasn't meant to be a direct visual, but we'd produced this sound of the 'event' so that the audience knew what was happening even though they couldn't see it. Later, when we considered the sound and the likely audience response, even though it made a lot of sense in the scene, we felt that too many people in the Costa Rican audience would find the sound offensive, so it was cut.

Arturo: Costa Rica

I would say that the amount of historical noise in a person's cultural environment can influence how they focus on the sound. It's like the cocktail party phenomenon, where you can focus in on one conversation in a room of noise. I think your environment, drawn from your culture, could affect how much you can focus on, say, listening to the actors even if there are many other audible sounds. At festivals for example, there is always a lot of background noise around the live performances. I think that, if an audience regularly interacts with sound in places where there's a lot of background noise, their attention and focus develop in a certain way that they don't notice the noise and can still enjoy the show. Without that previous experience, I think they would find the festival noise really distracting.

Angélica: Spain

I think there's a greater need to be accurate in the design of a soundscape that reflects a particular place if your audience is from that place. For example, if I was designing sound for a scene set in rural England near a train crossing, I would do my research and make sure that I was using the appropriate type of warning sound that plays in England when a train is approaching.

I remember, I once worked on an audio drama that involved a very tense, sexual scene. Coming from Venezuela, I grew up with tele-novelas, which are very expressive and passionate, so I felt I needed to make the audio scene for this project very passionate also. I used a lot of intense, dynamic sounds. There were even the sounds of objects being thrown around the space. I passed my work to the sound mixer. A little later I listened to the new mix they had created with my sounds. It had been heavily toned down, with many of my sounds much quieter in the mix. I had prioritised the passion in the scene. Others felt that the idea of romance was the element that needed the most emphasis.

Andreina: Venezuela

All things considered, it is absolutely the case that cultural factors can influence a person's perception of sound within a consumer context, be that a consumer of digital games, stage plays, film, or television (and these are of course not the only forms that could be explored). Whilst some felt that cultural factors and demographic did not, or should not, feature in the crafting of sound or soundscapes, contributors revealed numerous examples in which they have had to make adaptations. Some are very minor. Some are pretty substantial.

Chapter 7 summary: on the verge of global sound?

A handful of pages prior, the notion arose of a possible worldwide framework of cultural consideration for sound design. Now, reaching the end of our final chapter, we close with a couple of passages that pose a crucial question of whether we are moving towards a truly homogenous model of sound production, both in more objective terms such as compression and intelligibility, but also (and more worryingly if it transpires to be true) with regards to aesthetic and qualitative perceptions of sound.

> I think, as the world is becoming more globalised. The nationalistic silos of sound design and technique are beginning to fade.
>
> Arturo: Costa Rica

> I think that the games industry is becoming more globalised. Modern games need to reach an international audience to be financially successful. Companies are increasingly bringing in testers from a wider array of cultural backgrounds to get richer international feedback on their games. Obviously, the games industry has been global as a market for quite a while but, in terms of games production, now you have increasing numbers of games development companies springing up across smaller countries in Europe, across South America, and across Africa. Asia-based production is also growing rapidly. China is huge now in terms of games development. All the cultural backgrounds and influences of these developers, designers, artists, and companies is feeding into the global nature of the games being produced. I believe that sound design is absolutely being affected by that.
>
> Stafford: UK

> I notice that projects across different regions may choose to use their own, more native style, or they may sometimes choose to emulate the style of another region, but I think that the overall picture is very blurry with this emulating and blending of styles. I think sound design has become very global. As sound professionals, through the internet, we're exposed to so much content from around the world, but also more directly, through online tutorials. If we wanted to, for example, learn more about sound design in a traditional Japanese style, we can easily access and apply that information.
>
> Can: Turkey, China

Will our increasing exposure to content from around the world develop then reinforce a global standard? Will we slowly but surely limit our personal boundaries of what we determine as good and bad sound more and more? Personally, I don't believe so, and I look to a narrative drawn earlier by game sound designer Stafford Bawler: "I think that my personal quest for discovering things that are novel and interesting [...] goes back to my upbringing and the more unusual music that I was exposed to as a child". My hopeful interpretation of this statement is that, whilst future generations will unquestionably be exposed to avalanches of content from all corners of the globe, if that content is increasing in diversity, that will itself promote the desire for more diversity. The alternative is that the content increases in homogeneity, promoting instead the desire for more of the same. We are standing atop the mountain and may fall one of two ways. Whichever direction we fall in will generate a snowball effect. But which way will we fall?

Notes

1 The Wilhelm scream: https://www.youtube.com/watch?v=dc5F2C0CYlA (accessed 01.03.2022).
2 The Howie scream: https://www.youtube.com/watch?v=ZiWjZWQ3z0g (accessed 01.03.2022).
3 The Tartini effect is the illusory perception of a low musical note, perceived when only high notes are physically present.
4 The importance of good audio in online video content: https://vtrep.com/audio-is-more-important-than-video-picture-quality (accessed 01.03.2022).

References

Archibald, M. M., Ambagtsheer, R. C., Casey, M. G., & Lawless, M. (2019). Using Zoom Videoconferencing for Qualitative Data Collection: Perceptions and Experiences of Researchers and Participants. *International Journal of Qualitative Methods*, 18, 1609406919874596.

Augoyard, J. F., & Torgue, H. (Eds.). (2006). *Sonic Experience: A Guide to Everyday Sounds*. Montreal: McGill-Queen's Press-MQUP.

Barcelos, D. D., & Ataíde, S. G. D. (2014). Risk Analysis of Noise in Industry Making Clothes. *Revista CEFAC*, 16, 39–49.

Biot, M., & De Lorenzo, F. (2008). Some Notes on the Sound Reduction Index of Pax Cabins Panels on Cruise Ships. *Journal of the Acoustical Society of America*, 123(5), 3173.

Goodman, S. (2012). *Sonic Warfare: Sound, Affect, and the Ecology of Fear*. Cambridge: MIT Press.

Griffin, G. R., & McBride, D. K. (1986). *Multitask Performance: Predicting Success in Naval Aviation Primary Flight Training*. Naval Aerospace Medical Research Lab. Pensacola, USA.

Grimsley, L. (2020). How games have been altered in different countries. Culture of Gaming. Online article: https://cultureofgaming.com/how-games-have-been-altered-in-different-countries/ (accessed 05.02.2022).

Hayes, H. C. (1920). US Navy MV Type of Hydrophone as an Aid and Safeguard to Navigation. *Proceedings of the American Philosophical Society*, 59(5), 371–404.

Higashiguchi, A., & Shibuya, Y. (2021). A Study of Sound Presentation Effects on Silence During Video Conferencing. In Stephanidis, C., Antona, M. & Ntoa, S. (Eds.) *International Conference on Human-Computer Interaction* (pp. 566–570). Cham: Springer.

Imoto, K., Tonami, N., Koizumi, Y., Yasuda, M., Yamanishi, R., & Yamashita, Y. (2020, May). Sound Event Detection by Multitask Learning of Sound Events and Scenes with Soft Scene Labels. In Perez-Neira, A. & Mestre, X. (Eds.) *ICASSP 2020-2020 IEEE International Conference on Acoustics, Speech and Signal Processing (ICASSP)* (pp. 621–625). New York: IEEE.

Jiuliani, J. (2021). Major films that were changed for other countries. Looper. Online article: https://www.looper.com/94806/ (accessed 15.02.2022).

Leather, P., Beale, D., & Sullivan, L. (2003). Noise, Psychosocial Stress and Their Interaction in the Workplace. *Journal of Environmental Psychology*, 23(2), 213–222.

Loder, A. (2014). There's a Meadow Outside My Workplace: A Phenomenological Exploration of Aesthetics and Green Roofs in Chicago and Toronto. *Landscape and Urban Planning*, 126, 94–106.

Ma, K. W., Mak, C. M., & Wong, H. M. (2020). The Perceptual and Behavioral Influence on Dental Professionals from the Noise in Their Workplace. *Applied Acoustics*, 161, 107164.

Nakano, S. (2019). Sartorial choreography: the materiality and performance of clothes-making gestures (Doctoral dissertation, University of the Arts London).

Parker, J. E. (2019). Sonic Lawfare: On the Jurisprudence of Weaponised Sound. *Sound Studies*, 5(1), 72–96.

Portela, B. S., Constantini, A., Tartaruga, M. P., & Zannin, P. H. T. (2019). Sound Pressure Level in the Workplace: The Case of Physical Education Teachers. *Journal of Physical Education and Sport*, 19(2), 1153–1157.

Santhanam, S., Temesgen, S., Atalie, D., & Ashagre, G. (2019). Recycling of Cotton and Polyester Fibers to Produce Nonwoven Fabric for Functional Sound Absorption Material. *Journal of Natural Fibers*, 16(2), 300–306.

Talarico, M., Brancaleone, M. P., & Onate, J. A. (2020). Influence of a Multitask Paradigm on Motor and Cognitive Performance of Military and Law Enforcement Personnel: A Systematic Review. *Journal of Special Operations Medicine: A Peer Reviewed Journal for SOF Medical Professionals*, 20(1), 72–80.

Zollinger, S. A., & Brumm, H. (2011). The Lombard Effect. *Current Biology*, 21(16), R614–R615.

Conclusion

Let me start by saying, thank you for joining me on this voyage. To reiterate what will almost certainly feature in some of the marketing blurb, this book represents a year-long exploration into the human relationship with sound, emphasising the effects that cultural features can have on how we attribute meaning to sound, how we extract information from it, how we control it, how we feel about it and how it makes us feel, how it separates and connects us, and how we craft it and use it to create. It represents the collective insights drawn from conversations with 50 individuals, based across over 30 countries, and totalling roughly 120 hours' worth of discussion. You have already stuck with me through quite a substantial journey, and I won't keep you much further. We close with a very brief final consideration of the top-level questions that drove this project.

Final answers

Is the human relationship with sound different around the world?

Well, yes and no. Consistently across every overarching topic of discussion, contributors revealed many cases of both similarity and difference. In terms of sound and meaning, responses covered nearly the full range of theoretical perspectives, from the standard definition of sound as a soundwave, to the understandings that were deeply abstract, aesthetic, or subjective. Countering this variation is the observation that, across the world, *"what does sound mean to you?"* is a question many people have yet to ask themselves. Sound is undoubtedly the silver-medallist of the senses, the perpetual perceptual runner up to image, compared to which we are highly more likely to take for granted or, at the very least, utilise sound in more of a background capacity. This of course changes dramatically if you're either a sound professional or happen to live in the Antarctic.

Matters of culture again raise the issue of the relationship we have with sound not being something we pay much consideration to. In many cases, it appears as though we don't really think that much about how cultural factors influence our lives in any particular way, let alone when just talking about

DOI: 10.4324/9781003178705-9

sound. Despite this, the deeply rich and diverse nature of humanity, across lines drawn by seemingly any definition of culture, presents us with an unmistakeably dynamic and varied world of sound. Who we are and once were, where we are and once were, our experiences captured in memory, driving our actions in the present and our expectations for the future. No two people will be able to describe themselves in those terms and provide identical accounts – and as we have seen, all these differences in some way become manifest in sound.

Relationships between sound and place do not buck the trend established thus far. Across matters of sport, climate, transport, pets, wild animals, biodiversity, urban boundaries, and more, we consistently observed similarity, and we observed difference. The similarity is such that it is clearly possible to explain matters of acoustic ecology with models that possess a relatively manageable degree of complexity, but such models operate at a pretty high organisational level. The difference is in the fine print. Sound and place across the world could be described as broadly similar from a distance, but take a few steps closer, lean in, and the dramatic differences are revealed.

Lastly, and completing the set, sound and professional practice again reinforces the 'yes and no' answer on difference in our relationship with sound. Here if you remember, we ended on something of a cliff hanger. Numerous societal and technological factors, their impact upon global communication, and exposure to content are driving significant changes in how we experience and derive meaning from crafted sound, be it in film, television, music, theatre, or digital games. The practice of creating sound across these forms of media is certainly diverse, but the global communication factor has a meaningful impact here also, with designers, engineers, and other sound professionals now able to easily share techniques and approaches with peers all around the world.

At this precise moment in time, we are both similar and different. How the balance between these two will shift over as little as the next decade or so, I cannot say for sure but, to restate, I remain optimistic. This new power of communication and sharing opens doors to both greater diversity and greater homogeneity, but I believe that the predominant human drive is more towards embracing the former and not the latter. We seek to explore outwards more than to retreat inwards. We may indulge in nostalgia on occasion but, ultimately, we are inexorably drawn to the pursuit of progress.

What did you learn from your autoethnography?

The process of capturing the sound across various soundwalks was most definitely a positive experience and a good way of getting out of the house a bit more. Of course, the snippets of the reviews I wrote based on the soundwalks only represent a fraction of my notes and don't represent every location I captured, but I would argue that they do represent the broad features of my stream of consciousness as I considered the sound, then considered my consideration of the sound, and so on. The biggest accolade I can bestow upon this process was that it was a wonderful source of prompts. By that I mean, on numerous occasions, actively listening to what would otherwise be my everyday soundscape directed me to examine the questions we have observed throughout this book, that most people do not ask about themselves. Why did the soundscape of Old Portsmouth instil in me such a sense of home in spite of the fact that I have never lived there? Why do I have such an instinctive distaste for the sounds of perfectly happy people, revelling with a pint and indulging in the occasional cackle or guffaw? I'm afraid I'm not telling you the answers to these questions. But I did learn a few things about myself.

Do you still hate qualitative research?

Hate is a strong word. Let me re-phrase the question: *would you use qualitative research again?* Yes, but not in this way, not for love nor money. First the positive. The richness, intimacy, and personality of the information received were something quite unlike anything I have experienced throughout my research career so far. Numbers, whether they denote behaviours, scores, physiological measures, psychometric ratings, and so on, are cold. They most certainly have their benefits that run counter to the limitations (and grievances) of qualitative, but they have no life of their own. In comparison to the actual practice of producing and analysing the many transcriptions acquired for this project, processing quantitative data is sterile and metallic, the data itself evoking a sense of minimalism and indirectness. I honestly did leave every videoconference call with a smile on my face. It was a wonderful experience talking with each of the contributors and I genuinely never came away from a call feeling that the time had been wasted.

Now, the not so positive. Qualitative research, at least at this particular scale using this particular study design, is bloody exhausting. Honestly, I should have known better. I cannot recall precisely how many colleagues raised their eyebrows and emitted that thick-cheeked puff that implies "phew, that sounds like a lot of work, are you sure you're up to that?", but it was a lot. Each phase by itself was a substantial work-hour consumer, from setting up the interviews, to carrying them out, to transcribing, to coding (and re-bloody-coding) the key features, to pulling everything together. Add everything together and, let's put it this way, the next time a PhD tells me their workload is too intensive, I shall make sure to calmly but directly walk into the nearest empty office, shut the door, sit in a corner, and count backwards from 100. The alternative would be a disciplinary matter.

All things considered, I have absolutely no doubt that the methodology utilised was the right choice based on the overarching aims of the project. I still champion the value of qualitative data as a means of both enriching and cross-validating quantitative data within a mixed-methods approach and I feel that this experience has strengthened both my resolve and my ability to do so. Will I ever agree to spend over a hundred hours interviewing people, several hundred more transcribing their thoughts, and I don't know how long attempting to forge meaning out of it all? No. No I will not.

This is a message to you, future self:

Remember this moment? Do you really want to go through all this again? Why don't we go and wire up some undergraduates to an electroencephalograph, place them in total darkness, then subject them to a series of deeply unnerving sounds? That always cheers you up.

Thanks for reading,
Tom

Index